Canadian Construction and Maintenance Electrician

Certificate of Qualification
Exam Preparation, 2008

Tony Fazzari

Centennial College Press

Library and Archives Canada Cataloguing in Publication

Fazzari, Tony

Canadian construction and maintenance electrician : certificate of qualification exam preparation / Tony Fazzari.

(Centennial exam prep series)
ISBN 978-0-919852-62-4

1. Electric engineering--Examinations--Study guides. 2. Electric engineering--Examinations, questions, etc. 3. Electricians--Certification--Canada. I. Title. II. Series.

TK169.F39 2008   621.3'076   C2008-903544-5

Copyright © 2008 by Tony Fazzari

ISBN: 978-0-919852-62-4

Centennial College Press
951 Carlaw Avenue
Toronto, Ontario
M4K 3M2
www.centennialcollegepress.com

All rights reserved. This publication is protected by copyright and permission should be obtained from the publisher prior to any reproduction, storage in a retrieval system, or transmission in any form or by any means—electronic, mechanical, or otherwise.

The publisher and author would like to thank Sean Bennett and Dave Weatherhead for generously giving their permission to adapt, and in the case of Chapter 2 (excepting minor adaptions by the author) and some sections of Chapter 1 and Chapter 3, to reproduce word for word, Sean Bennett's explanations of C of Q testing and exam-writing strategies. The material in these sections, excepting the author's changes, omissions, or adaptations, is owned by Sean Bennett and Dave Weatherhead, and was originally published in *Canadian Automotive Service Technician C of Q Test Preparation, 2nd Edition*, copyright © Sean Bennett and Dave Weatherhead, 2007.

The publisher and author would also like to thank Fred Diamanti, Program Coordinator, The Ontario Ministry of Training, Colleges and Universities, for his review of Chapter 1 and Chapter 3.

Content review and technical editing by Jim Gamble

Book design and typesetting by Computer Composition of Canada Inc.

Printed in Canada by Webcom

Contents

Chapter 1	Introduction 1	

- How This Book Prepares You
- Things You Need To Know about the Construction and Maintenance Electrician Red Seal Exam
- Using This Book

Chapter 2	Exam Strategies 5	

- Review 1
- Review 2
- Review 3
- Review 4
- Active Learning
- Reading Textbooks
- Improving Retention
- Improving Exam-Writing Skills

Chapter 3	Using Exam Blocks 13	

- The Exam Blocks
- Study Smart Using NOA Tasks and Sub-Tasks
- Education, Job Skills, and the C of Q Exam
- Preparing for Provincial C of Q Exams
- Key Points

Chapter 4	Learning to Use the Canadian Electrical Code Book 21	

- Different Search Methods
- Tips
- Practice Questions: Using the CEC Book
- Answers to Questions

Chapter 5	Practice Exam 1 31
Chapter 6	Practice Exam 2 49

Chapter 7	Answers to Practice Exams 69	
	• Answer Key to Practice Exam 1	
	• Answer Key to Practice Exam 2	
Chapter 8	Explanations for Practice Exam 1 Answers 71	
Chapter 9	Explanations for Practice Exam 2 Answers 91	
Chapter 10	Recommended Study Texts 109	

Introduction

This book is a exam preparation guide for Canadian Red Seal Certificate of Qualification testing (C of Q) in the Construction and Maintenance Electrician trade. Apprentices can also use this book to prepare for any provincial C of Q exam (all certification exams for construction electricians in Canada are similar). This resource will not replace the hands-on experience of a 4- or 5-year apprenticeship, the in-school formal theoretical learning period, or textbooks that deal in detail with the subject matter. Our goal is to help students who are in a final preparation phase before writing a Certificate of Qualification exam by helping them

- understand the type of questions on the exam and see what it is like to write one
- take advantage of knowing the relative weighting of the various subjects
- use the best methods to study for and pass the C of Q
- practise writing the exam before they take it by answering 2 complete practice exams
- build confidence and identify the areas they need to improve on before taking the C of Q

This book will be excellent preparation for the C of Q exam.

How This Book Prepares You

The first 4 chapters explain what the Red Seal exam is, how it relates to provincial exams, and how best to prepare for a Red Seal or provincial exam. Chapter 1 gives an overview of how the C of Q exams are created and what kind of knowledge they focus on. Chapter 2 presents advice on how best to study for the exam and tips on writing it. Chapter 3 explains how to use your knowledge of the exam profile (the number of questions each exam subject gets) to decide how best to weight your study time, and also shows how the exam focuses on practical knowledge and skills and not academic knowledge. Chapter 4 shows how to use the Canadian Electrical Code book to locate important material quickly to speed up studying.

The largest section of this book contains the 2 practice exams and answers. Each exam contained in Chapters 5 and 6 imitates the Red Seal exam. Each is a complete practice exam of 100 questions written in the Red Seal format. While the questions in each are not identical to those you will write on the exam, they are very similar to them. In many cases they are more challenging, and understanding any of these questions and their answers (explanations for answers are given as well) should mean apprentices will know more than enough to answer any C of Q question on the same topic. Chapters 7, 8, and 9 give the answers to the practice exams and also explain the answers thoroughly.

Things You Need to Know about the Construction and Maintenance Electrician Red Seal Exam

Red Seal Construction and Maintenance Electrician exams are national Certificate of Qualification exams that qualify an apprentice who passes them to practise their trade in any province or territory in Canada. The Red Seal exam, like any provincial C of Q exam, tests *practical* knowledge, skill, and competence. In fact, these practical questions, Red Seal or provincial, are all ultimately drawn from the same source: the National Occupational Analysis (NOA), a federal document that defines all of the skill areas of a specific trade in which a tradesperson in Canada must prove their competence in order to be licensed. For this reason, the practice exams will prepare an apprentice equally well for any C of Q exam in any province or territory.

How the Questions Are Written, and Who Writes Them

Red Seal trades are sponsored by the federal government, who assigns each of the Red Seal trades to one host province that administers and creates a Red Seal exam for the entire country. The current host province for the Construction and Maintenance Electrician trade is Nova Scotia. Representatives of the host province invite resource people, all qualified trade practitioners, from across all provinces and territories to participate in workshops to create the exam.

Once the exam has been developed, it undergoes a peer review performed by qualified electricians from each of the participating provinces. The peer reviewers offer feedback to help ensure that the questions of the exam constitute an up-to-date and valid means to evaluate an apprentice's proficiency. Once this feedback has been addressed, the finalized exam is implemented as the new Red Seal examination.

The workshops create the exam questions based upon National Occupational Analysis *Sub-tasks*, the most specific categories of skills into which the NOA divides a given trade. Chapter 3 will show you how this works, and how to use your knowledge of Sub-tasks to prepare for the exam. Arguably, you can gain an even greater advantage by planning your studying around the weighting of the Red Seal *Exam Blocks*, discussed below.

Why Exam Blocks Are Important

The Red Seal exams are created to mirror the main skill areas of the NOA, called *Blocks*, which break down into more specific skill areas called *Tasks*, which in turn break down into the Sub-tasks. NOA Blocks are important because the Red Seal exam questions test apprentices' *practical* knowledge, and the Sub-tasks that make up each Block are mainly concerned with that practical, or hands-on knowledge, as opposed to the academic knowledge gained in school. In fact, C of Q workshop writers refer to the individual Sub-tasks to develop each question to test an apprentice's competency in a specific skill. As such, looking at Sub-tasks is like a tour of all the very specific topics you could be tested on.

> *By knowing the NOA Blocks, you will be able to focus your study on the right things to pass the exam—the hands-on rather than the academic.*

But there is another advantage. The NOA Blocks also form the major sections—or "Exam Blocks" —of the Red Seal exam. The exam headings, or Exam Blocks, match the Block headings of the most current NOA for the trade. So, by looking at the breakdown of the NOA Blocks (see Chapter 3), you can quickly see how many exam questions each Block is worth. You will then be able to decide how much of your available study time to devote to each Block, and not waste time studying hard for areas that get very few questions.

For instance, in Chapter 3, you will note that the Block *Distribution and Services* has 24 questions dedicated to it, but the Block *Extra Low Voltage Systems* has only 10. This means that every Red Seal C of Q will have exactly that number of questions on each of those subjects.

Knowing the number of questions for each Block is important when it comes to studying for an exam because it means you can prioritize study time by the weighting—that is, the relative importance of each subject. Knowing that there are 24 questions on *Branch Circuit Wiring* (heavily weighted)—or 24% of the exam—you should probably devote roughly 24% of your available study time to this subject.

C of Q Exam Questions Test Practical Knowledge, Not Academic Knowledge

Because both the NOA and any province's Training Standards both focus on the practical knowledge and skills of the trade, each system reflects the kind of knowledge you will be tested on much more than your college curriculum. Therefor, you will perform best on the exam if you review the individual skills listed in either the NOA Sub-tasks or the *Performance Outcomes* of a provincial exam's Training Standards (see Chapter 3), and then arrange to practise these skills as well as to devote some time to review your course work from school.

Again, because C of Q questions are ultimately all sourced from the National Occupational Analysis (even provincial Training Standards are created from the NOA) and NOT from the in-school provincial curriculum you studied in college, it is therefor possible that you will have questions on procedures and technologies you did not study in the college curriculum. Provinces and territories design their own college curriculum, and this curriculum tends to be changed more frequently than the Training Standards or NOA.

Using This Book

The questions in the exams of this book have been constructed to broadly cover the Sub-tasks identified in the National Occupational Analysis of the Construction and Maintenance Electrician trade to help the student prepare for the C of Q.

This is an exam preparation book, not a comprehensive textbook. It is designed to complement and reinforce knowledge you already have. If you are unable to understand an explanation, you may have to refer to some of the textbooks recommended at the end of the book. If you have completed the required in-school training, you are probably in possession of one of the comprehensive electrical textbooks, and this is all you will need. If you have a specific weakness in a subject area, you may benefit from one of the specialty textbooks. These will often go into a technology with a thoroughness well beyond that required for a Red Seal certification examination, but the key to correctly answering an exam question is a thorough understanding of the subject.

Exam Strategies 2

Every student learns differently, so the learning method that works best for one person may not work for another. Generally, you should not leave review and study time to the last moment. Spending 15 minutes a day for 6 months will be more effective for most learners than spending a week before an exam studying to the point of exhaustion. Most learners forget material if they do not review the subject matter. Look at the following data produced from testing following a lecture:

% Retained in Memory	Without Review	With Review
After 24 Hours	60%	80%
After 1 Week	50%	75%
After 2 Weeks	40%	70%
After 3 Weeks	30%	70%
After 4 Weeks	20%	70%

National Institute for Staff and Organizational Development

The above chart shows the value of establishing a routine of review. For most learners, reviewing material is a key to succeeding in exams. The great thing about making a review a habit is that it does not take long. If you properly understood the material first time around, minutes are all that is required to refresh the learning experience in your mind. The following is an example of a review timetable that will work for most students:

Review 1

This should take place within 24 hours of the learning experience. The same day works best for most learners. This review step is the most important in the review process. Use point-form notes—that is, keep notes as brief as possible. When instructors write out notes or issue handouts, abbreviate them. Short-form notes are a great study tool; just make sure that they are not abbreviated to the extent that they cease to have meaning to you. In this first review, try to understand the subject matter. Make notes about material you do not grasp and make it your business to get those questions answered by either private research or your teacher.

Review 2

Go over your abbreviated notes in this session. Make sure you understand both your own notes and the subject matter covered. This second review should take a matter of minutes; you will be recalling the learning experience and the time you devoted to the first review.

Review 3

In this step, you simply relive the initial learning experience and the first review step. It should take little time and you may even feel it is boring. However, it is a great way to reinforce the information learned. Depending on what type of learner you are, you may want to do this once a month. But do it, it does work, and the payoff will come when you are tested.

Review 4

This should be done immediately before testing. Most learners do not retain information well when their only study occurs immediately prior to an exam. Try not to spend too long studying because if you do, your brain will be exhausted by the time you have to challenge the exam. One thing that can work well is joining a small study group before an exam. Use sets of typical exam questions such as those found in this book as a basis for discussion.

Active Learning

Construction electricians tend to learn best by doing rather than thinking, which is why we often feel out of place in classrooms. Become active in the classroom by using some of the following strategies:

1. *Make your own notes.* Most teachers write far too much on boards and in their handouts, so rewrite what they say in terms that mean something to you. Notes are for you only. Challenge yourself to make them as meaningful as possible. Remember, notes are a great review

instrument, especially if you can capture the essence of a 3-hour class, in 1 page of bullet-form notes.

2. *Draw diagrams.* Try not to always rely on handouts. Drawing a diagram makes you active in the class. Even if you don't have much artistic ability, the actions required to draw a diagram will usually help you better understand the technology.

3. *Ask questions in class.* That is, ask them if you feel comfortable doing so. Not all construction electricians like to ask questions in class. But asking questions, even if you think they are dumb questions, is a way of making yourself an active learner, and that helps retention.

4. *Look for ways of making connections between the theoretical information you get in class and the hands-on practice of construction technology.* This works well if you have worked hands-on with the technology you are learning in class.

Reading Textbooks

Most textbooks are difficult to pick up and read in serial fashion like a novel. When you use a textbook to study, define what your goals are before opening the book and then use the book to meet them. Before you begin, it is a good idea to know something about the goals you want to achieve, so write them down in brief form. Next, you have to navigate the contents of the textbook to achieve those goals. Here is an action plan I use when studying subject matter that is new to me:

1. *Goals.* Define your goals. Write them down on paper. This could be as simple as a couple of words or it could be more complex, depending on what it is you are expecting to learn.

2. *Survey.* You have your goals in note form on paper. Next, consult the textbook table of contents to determine if the book is going to be of any help. Once you have targeted a chapter in the textbook, take a look at the chapter objectives: this should give you a pretty good idea of whether the chapter is the best to start with.

3. *Read, distill, and restate.* Target the information you need in the text. Read the content and get a general sense of what the author is trying to say. In most cases, what the author is saying can be restated in fewer words. Using pen and paper, take bullet-form notes on the content. Keep it brief. Most important: put everything in words that make sense to you.

4. *Review.* Close the textbook and go over the bullet-form notes you have

made. Do they make sense to you? If they do not, you will have to open the book up again and redo the notes. If it does make sense and the notes you made are sufficient to allow you to recall the material later on, you have achieved your objective.

Improving Retention

Because we all learn differently, remember that what works for you may be different from what works for others. We said before that construction electricians generally learn best when a learning experience is active. That very often means hands-on learning that is not always possible in college programs with congested curriculum and large classes, so look for other ways of making the learning experience real for you. Here are some key things that might work:

- *Active classroom learning.* Use the classroom techniques we described earlier to avoid being a completely passive participant in the learning process. Because you are more likely to learn by doing, DO as much as possible in class. Condense existing notes, draw diagrams, and ask questions.

- *Research information.* Use the Internet if this approach to learning works for you. It is a great way of supplementing what you have learned in the classroom, and the best thing is that you control both the learning path and the pace. Again, make short bullet-form notes when you are hunting down information online.

- *Use video.* Most colleges have enormous resource centres full of videos and DVDs too lengthy to run in structured classes. If you like to learn visually, use the video, CD, and DVD libraries in resource centers to learn. Note: video is a big turn-off for some and has a way of inducing sleep. If this is true for you, recognize it and avoid long video presentations.

- *Active research.* This can be really effective. In teaching construction technology, it is common to hear students complain that they have never worked with this type of equipment and find it difficult to understand some applications. If you find you are not comfortable with an aspect of the trade, such as residential wiring, for a small investment you can purchase a few device boxes, some NMD90 (romex) wire, a few receptacles, 2 three-way switches, and 1 four-way switch. Make a project for yourself, such as wiring 1 or 2 light fixtures that are to be controlled by the 2 three-way and 1 four-way switches. Draw a diagram of the circuit or a number of diagrams showing different ways of wiring the circuit. Last, review the code rules and complete the job. What you learn by practical experiences like this one will stay with you for a long time.

- *Be curious.* This is your career. Don't black box technology. If you do not understand how a particular component functions, get one that has failed, test it with test instruments, and take it apart, with a hacksaw if necessary. Make it your business to answer those questions that no one seems to have an answer for.

Improving Exam-Writing Skills

Most construction electricians are not academics by inclination, and it often seems unfair that they are examined academically rather than by practical tasks. The modern construction electrician must be literate, so attempting to justify a paper examination failure by claiming to have mastered all the hands-on competencies fools no one but you. However, there are some simple things you should be aware of that can greatly improve your ability to succeed in exams:

1. *The number one reason for failing an exam is simply not knowing the material.* Use the study techniques described in this guide to ensure that you understand the subject matter.

2. *Never spend too much time analyzing an exam question.* If you do not understand a question, skip it and return to it when you have completed the questions you do understand. Analyzing is a great skill for an electrician—but it can hinder you when writing an exam by causing you to read meanings into an exam question that are not there.

3. *A typical exam question consists of a stem or question followed by 4 possible answers.* Only 1 of the 4 possible answers is correct; the other 3 are known as distracters. When you are not sure of the answer, try eliminating obvious distracters by crossing them out. This will help narrow down your choices.

4. *Read with your pencil.* Underline key words in the question. Eliminate distracters that make no sense at all. If you have to guess, make sure your guess is an intelligent one—you can only do this if you have already eliminated the distracters that do not make sense.

5. *Read the question and* ALL *the answers.* In a multiple choice exam you are selecting the most correct answer; it may be that some of the distracters are in some way correct.

6. *Distribute your time appropriately.* This is especially important if you know you write exams slowly. Take off your watch and place it above the exam paper. Divide the exam into sections.

7. *Forget about answer patterns.* If you have answered question A for the previous 4 questions it is of no significance.

8. *Think twice before changing an answer in a multiple choice exam.* Studies show that more often than not a correct answer is changed to an *incorrect* answer—take a look at bullet #2 on the previous page about over-analyzing exam questions. The answer that first occurs to you is likely to be correct.

9. *Erasing.* Certificate of Qualification exams are graded optically by scanning. If you erase an answer, make sure that it is completely erased. If not, 2 answers will be scanned on the question and you will get the question wrong. If you are unable to properly erase a response that you feel is incorrect, ask for another answer sheet.

10. *Answer every question.* In a C of Q exam there are 4 possible answers for every question and only 1 answer is correct. You do not get penalized for incorrect responses. Even if you are outright guessing the answer, you still have a 25% chance of getting it right!

11. *Relax!* If the exam becomes confusing to you, spend 5 minutes daydreaming to get your mind off the exam. You may just find that what was confusing will become less so. Breathe slowly and deeply if you are inclined to panic during exams.

12. *Don't expect every question in the exam to be absolutely technically accurate.* Certificate of Qualification exam questions are written by teams of journey person electricians, not academics. Despite checks and a proofing procedure, it is possible for the odd question to appear in an exam that makes little sense. If you get a question like this, try asking yourself which answer the author of the question might have thought was correct. More important, avoid getting too worked up over one bad question! Move on to the next.

13. *In any question involving personal or shop safety, think about the correct way, not what you might think is accepted practice.*

14. *Use 2 basic strategies when writing a test.* 1) Read through the exam and answer the easy questions first. Studies have shown that a few

difficult questions at the beginning of an exam can put the writer in a defeated mood; the easy questions now seem harder, affecting their performance and lowering their overall grade. Answering the easy ones first may keep you in a more positive mood. 2) Divide up your time according to the number of questions that you have, and then spend the appropriate amount of time on each question. If you are unsure about an answer, shade it in lightly, mark the number of the question on a separate paper, and go back to it if time permits. With a scan sheet and 100 questions, it can be easy to fill in the wrong space, and realizing this late in the exam can be devastating.

15. *Be organized and stay calm.* When the time comes to start the exam, you will find a test, a scan or answer sheet, a booklet of drawings, a calculator, a Canadian Electrical Code book, and a pencil in front of you. This can be overwhelming, so it is very important to be organized and stay calm.

Using Exam Blocks 3

In this chapter we will look at the divisions of the National Occupational Analysis (NOA) and the weighting of each one. Because these are the same divisions of the C of Q exam, by making yourself familiar with these divisions you will be able to decide how much study time you need to devote to each subject on the exam to perform well. Furthermore, we will also discuss the value of using the Sub-tasks, from which all Red Seal and perhaps all provincial exam questions are sourced. Last, we will look at a hypothetical example of a C of Q question in order to demonstrate the kind of practical knowledge they demand.

The Exam Blocks

The Exam Blocks are essential, because they make up the exam profile. Again, the term *exam profile* simply means the groupings of the exam questions into main headings, or "Blocks," and the number of questions each heading gets. The NOA looks at the competencies required of a Construction and Maintenance Electrician and divides these competencies in the following manner:

Block: A Block is the largest division of skills within the analysis and can be regarded as a major set of skills sharing at least one thing in common. For instance, *Branch Circuit Wiring* is the title of one of the 6 Blocks that compose the NOA. These 6 Blocks are important in the exam profile, because each identifies a reportable subject for purposes of constructing the C of Q exam.

Task: A Task is a distinct activity that falls under a Block of skills. For instance, *Installs Raceway Systems and Cables* is a set of skills that fall under the Block *Branch Circuit Wiring*.

Sub-task: A Sub-task is a specific competency or skill that falls under a Task. For instance, *Installs Boxes, Cabinets, and Fixtures* is a specific skill that would be one of many required to achieve competency in the Block *Branch Circuit Wiring*.

Construction and Maintenance Electrician Exam Blocks

Questions within the Certificate of Qualification are organized by these Exam Blocks:

Block A: Occupational Skills
Block B: Distribution and Services
Block C: Branch Circuit Wiring
Block D: Motor and Control Systems
Block E: Extra Low Voltage Systems
Block F: Maintenance, Upgrading, and Repair

Following is the Red Seal exam profile.

	BLOCK	**% of Exam**	**# of Questions**
A.	Occupational Skills	12%	12
B.	Distribution and Services	24%	24
C.	Branch Circuit Wiring	24%	24
D.	Motor and Control Systems	19%	19
E.	Extra Low Voltage Systems	10%	10
F.	Maintenance, Upgrading, and Repair	11%	11
	TOTAL	100%	**100**

Even at this broadest division of skills, there will likely be areas of strength for you and perhaps one or two areas in which you could be better prepared. You now know how many questions will be aimed at each area. You can take advantage of this now and speak to your instructor or apprenticeship supervisor about getting some extra time in the shop to practise the skills that fall under the areas you want to strengthen. Naturally, you should also plan to review all the relevant in-school material to reinforce your hands-on study.

Exam Block Weightings

We have seen how each of the 6 Exam Blocks is weighted. This weighting does not change from exam to exam. For instance, if you and a friend go to write your C of Q on the same day, you will not necessarily write exactly the same questions, but you will both write exactly 19 questions relating to *Motors and Controls*. This is absolutely reliable for Red Seal C of Qs, but you can apply the same strategy even if you are writing a provincial C of Q, as we will discuss shortly.

Whether you will be writing a Red Seal or provincial exam, begin preparing to write your C of Q by taking a close look at the weighting of each Exam Block. This will help you use your available study time strategically to review as much as you can in the areas that count the most, or compensate best for your weakest areas. If you check the chart above, you will notice that there are 24 questions on *Service and Distribution* and only 10 on *Low Voltage Systems*. Assuming you were equally confident in all of these subjects, it is obvious that you should dedicate more time to the study of *Service and Distribution* when prepping for your exam.

Study Smart Using NOA Tasks and Sub-Tasks

This section lists the Tasks that make up the NOA Blocks for the Construction and Maintenance Electrician exam, and gives the percentage of each Block that each Task makes up in the NOA. We will not list the Sub-tasks; there are over 1000 of them. However, you can use this breakdown of the Tasks to further tailor your preparation. For example, you might have worried that you needed to prepare for everything under the Block *Branch Circuit Wiring*, but when looking over the Tasks and the weighting of each, you realized you really only feel weak in Task 10, *Installs Raceway Systems*. Once you know this, you can stop worrying as much about the other Tasks, and then pull up the 10 to 20 Sub-tasks that fall under *Installs Raceway Systems*.

Although not every Sub-task is the basis for an exam question, every question on the exam is based completely upon one specific Sub-task. This means that when you look at a Sub-task, you are looking at the same defined topic a C of Q writer used to craft a question—so you should be able to prepare for specific Tasks or even Sub-tasks very efficiently, practising only what you need to in order to fill in gaps in your knowledge.

Each of the following 20 Tasks are shown with the weighting of each Task within its Block. To view the whole document, go to the Red Seal Web site (www.red-seal.ca) and find the NOA for Construction and Maintenance Electrician.

Construction and Maintenance Electrician Blocks and Tasks

Block A: Occupational Skills

Task 1: Interprets Occupational Documentation. (32%)

Task 2: Organizes Work. (23%)

Task 3: Communicates in the Workplace. (18%)

Task 4: Uses and Maintains Tools and Equipment. (27%)

Block B: Distribution and Services

Task 5: Installs Service Entrance. (28%)

Task 6: Installs Sub-panels, Feeders, and Transformers. (27%)

Task 7: Installs Bonding, Grounding, and Cathodic Protection Systems. (22%)

Task 8: Installs Power Generation Systems. (13%)

Task 9: Installs High Voltage Systems. (10%)

Block C: Branch Circuit Wiring

Task 10: Installs Raceway Systems and Cables. (36%)

Task 11: Installs Power and Lighting Systems. (31%)

Task 12: Installs Heating and Cooling Systems. (19%)

Task 13: Installs Emergency Lighting Systems. (14%)

Block D: Motor and Control Systems

Task 14: Installs Motor Controls. (66%)

Task 15: Installs Motors. (34%)

Block E: Extra Low Voltage Systems

Task 16: Installs Signaling Systems. (53%)

Task 17: Installs Voice and Data Systems. (47%)

Block F: Upgrading, Maintenance, and Repair

Task 18: Upgrades Electrical Systems. (43%)

Task 19: Maintains Electrical Systems. (35%)

Task 20: Performs Preventative Maintenance. (22%)

Education, Job Skills, and the C of Q Exam

Most apprentices eligible to write a Canadian Red Seal Certificate of Qualification will have attended the in-school training sessions delivered by provincial colleges and managed by provincial education authorities. The formal curriculum used in the in-school portion of apprenticeship training is designed to complement the content of the Training Standards. Theoretical knowledge is a key building block to any technical skill set, so the curriculum emphasizes plenty of theory en route to achieving competency. But remember this: **the federal exam you are going to write is based not on the curriculum you studied in your college program, but directly on the National Occupational Analysis**.

Remember also that **the Certificate of Qualification exam attempts to make a determination of your *competence* in each task, not just your ability to remember facts and procedures**. This means that you should not see questions that are purely theoretical. Each question is going to involve a *practical* application. This does not mean you do not have to know any theory—but it does mean you will be unlikely to see a question like this typical in-school example in a C of Q exam:

What is the total capacitance of 3 capacitors connected in series if C1 has a capacitance of 20µf, C2 has a capacitance of 30µf, and C3 has a capacitance of 60µf?

You will see this type of question in tests when you study electrical theory in college, because the curriculum has a mandate to look at technology from a more theoretical point of view.

In a Red Seal C of Q exam, you are more likely to see a question like this:

What would the current be on the neutral conductor of a 120V/240V circuit, when L1 is determined to have 2800 W and L2 is determined to have 1300 W?

    A.   0 A.

    B.   6.25 A.

    C.   9 A.

    D.   12.5 A.

The correct answer is D.

In this question, your knowledge of an on-the-job troubleshooting procedure is being tested together with your academic competency with theoretical concepts.

One thing that cannot be emphasized strongly enough is the importance of understanding the role of college in-school training and that of workplace training when strategizing your study preparation. **In terms of total apprenticeship time, typically 10% is spent in formal in-school instruction, while 90% is on**

the job. Not everything you need to know is going to be learned in the in-school portion of apprenticeship. On-the-job training is fundamental to apprenticeship; in fact, it is what makes it effective. So expect questions that test your on-the-job knowledge.

Preparing for Provincial C of Q Exams

Provincial exams are developed and administered by the individual province, and so any separate provincial exam may have superficial differences from the Red Seal exam, or from a different provincial exam. For example, the exam as a whole could have more or fewer questions, with different numbers of questions in each Exam Block. That said, all provinces are included in the Red Seal program and offer the exam, and at least 1 province, Ontario, has abolished its provincial exam entirely and requires all Ontario apprentices to pass the Red Seal.

Another likely possibility is that your province may use the same Red Seal exam, but only insist upon a 60% score for a pass, and on a score of 70% or higher will award an *interprovincial* license. We advise you to ask your apprenticeship councillor whether you will write a provincial exam or a Red Seal exam (and remember to ask *which* exam you write, not just whether or not you need 60% or 70% to pass), what source material it is based upon, and how you can get a copy of the exam profile. As long as you follow these steps, you can apply all the strategies in these opening chapters to any provincial exam.

Training Standards and National Occupational Analysis

In most provinces, when you sign an apprenticeship contract you are issued a Training Standards booklet. Your supervisor is expected to sign off on each of the Training Standards as you progress through the apprenticeship term. Although it is important to note that the Red Seal Certificate of Qualification examination is based on the National Occupational Analysis, it is still valuable to refresh your memory of the subject areas using the Training Standards. They too identify the skills you will be tested on, only using a somewhat different breakdown of those knowledge areas. By focusing on the practical knowledge described by either the NOA or the Training Standards, you are making sure NOT to focus too much on the academic knowledge gained in college.

As we have said, the questions for Red Seal exams are sourced from the Sub-tasks. **Any separately created provincial exam might well do the same, but if not, it would likely use that province's Training Standards as a source of exam questions.**

To see how similar the Training Standards are to the NOA, you need only compare them. The major divisions in the Training Standards are the *Skill Sets*, which correspond with the NOA Blocks. The next division down is the *Performance Outcomes* (on-the-job training outcomes), which are similar to the Sub-tasks of the NOA—in fact, they are derived from them. At this level, the two systems basically meet. Performance Outcomes are on-the-job training statements that expand on the Sub-tasks listed within the NOA, not only listing the Sub-task, but stating how the apprentice is to perform it, with what tools, and under what safety standards and government regulations. Furthermore, the Red Seal administrators insist that all provinces and territories that wish to be governed by its standards must have at least 70% correspondence between their Training Standards Performance Outcomes and the NOA Sub-tasks for each trade.

The practical point is that at the level at which exam questions are sourced for the Red Seal (Sub-tasks), and at the level they will almost certainly be sourced for any provincial exam (Performance Outcomes), the two are very alike, having only slightly different names, and perhaps slightly different Exam Block weightings. Ask your apprenticeship counsellor to confirm what level of Training Standards is used to source the questions, and then use these to plan your in-shop practice and coursework study.

The practice exams will prepare you very well for *any* exam questions on any C of Q exam, but to use the strategic advice in the first chapters, be sure you know what exam you are writing (if not the Red Seal), and how each Exam Block is weighted.

Key Points

- Exam questions relate to the National Occupational Analysis Sub-tasks, or the Training Standards Performance Outcomes, both of which are about hands-on knowledge and skill. Expect lots of questions that relate to hands-on procedures.

- Strategize your exam preparation by looking at the number of questions dedicated to each Exam Block.

- Apprenticeship consists of approximately 10% in-school training and 90% on-the-job training. The C of Q is designed to determine your **on-the-job competence**, not just the more theoretical knowledge you learned in the college curriculum.

Learning to Use the Canadian Electrical Code Book  4

In order to do well when writing the C of Q, an understanding of the Canadian Electrical Code (CEC) book and how it organizes important information is essential. The C of Q is based largely on your apprenticeship training, but to answer as many as 40 to 45 questions, you will need to quickly locate information found in the code book.

Different Search Methods

There are different ways to search for the information you need, and although the CEC book is well organized, you have to think in order to make your search as efficient as possible. Here are the methods.

Using the Index

Step #1: Look for the main word in the question that defines the topic, for example: *conductor* or *conduit*.

Step #2: Look for the secondary word that describes what you want to know about the main word, for example: *conductor ampacity* or *conduit, minimum size of*.

Step #3: Use these words to find the appropriate rule to answer the question. If this is unsuccessful, move on to Step #4.

Step #4: Change the *order* of the main and secondary words.

Step #5: If it appears the words you are using are not leading to the information you need, try a new main and secondary word that might be better targeted.

Step #6: Change the main and secondary words to words that are recognized by the code book. Here are some examples:

fixture = luminare; wire = conductor; switch = disconnecting means

Step #7: When the index gives more than 1 rule, write them on scrap paper, and look at the first 1 or 2 digits to see the section number. This should give you an idea of whether or not you are looking in the right place. Before plunging into each rule, scan the title first to see if the rule seems likely to contain the information you need.

Here's a simple example that uses only the first 3 steps:

Question: What is the maximum rating of a plug fuse?

The main word is the basic concept you need to know something about, in this case, it's *fuse*.

The secondary word, or words, of the question define what we want to know about the main word. In this case, it's the word *rating*.

Using the index of the code book, searching under *fuses, rating of* takes you to Rule 14-208. Once you have found the specific rule, the answer is usually not hard to find.

Answer: 30 A.

Using Appendix B

If you feel you are looking at the correct rule and cannot find the answer, chances are that the answer is in Appendix B, a special section of supplemental notes on the rules.

An example:

Question: If 30 m of 53mm of polyvinyl chloride (PVC) conduit is placed in a temperature range of 20°C–25°C, how much expansion is likely to occur?

Looking in the index under *conduit, expansion and contraction of*, you will find Rule 12-1118, which explains that the maximum amount of expansion in a run of PVC conduit before an expansion joint is required is 45 mm. This did not answer the question, but the rule does tell you to check Appendix B. The section for Rule 12-1118 in Appendix B not only gives you a coefficient of expansion for PVC but also gives you a formula and a worked-out example.

Expansion = length of run in metres × temperature change × coefficient of expansion.

Ex = 30 m × 45°C × 0.052

Answer: Ex = 70.2 mm

Using the Table of Contents

A number of rules do not appear in the index of the CEC book. Some listings are for the section only and do not list rules. For this reason, the Table of Contents can be a useful tool.

Example:

Question: Where should the leakage current collector for a hot tub be installed?

In the index, the only reference made is to *pools tubs and spas, Section 68*. Scanning the Table of Contents, you will find under Section 68 the sub-heading Spas and Hot Tubs, page 252. Upon turning to page 252 you find the heading Spas and Hot Tubs, and then are able to quickly scan down to Rule 68-406, Sub-rule 1, which gives your answer.

Answer: In all water inlets and outlets.

Using the Tables

In some cases, the answer to a question is more likely to appear in a table than in a rule. Just after chapters on the individual sections of the CEC book, you will find a division of the book for tables. This division of the book begins with an index on the first page which you can use to quickly locate information.

Example:

Question: What is the full-load current of a 10hp 230V three-phase motor?

Using the index for the tables, you can scan the headings of tables to quickly find Table 44: Three-Phase AC Motors, which contains the answer.

Answer: 28 A.

Appendix D

Appendix D contains useful information, often in table form, and like the Tables section we just discussed, it has an index on the first page. It might be useful to know that the tables located in Appendix D need to be updated more frequently than those in the Tables section, but otherwise, there is no rule of thumb to tell you whether a specific piece of information would be more likely to be in Appendix D than in the Tables.

Therefor, it is important to be able to quickly navigate to the information contained in *both* places before writing your C of Q.

Tips

Remember, these methods still require you to think about how best to get the information, and sometimes to come up with a different approach. Here are some tips, and some advice on getting the most out of the CEC book.

Scanning

Sometimes all of these suggestions may fail. As a last resort, pick the section of the CEC book that you think is most likely to answer the question and scan it for the obscure rule. Remember, to do this effectively, you are reading *strategically*, which means looking for key words in headings first, then in passages of text—but don't read line for line hoping to stumble onto the information you need. That takes much too long, and you are more likely to find what you need by letting your eye move fast, not only left to right, but also up and down the page.

Exceptions to Rules

When answering a question, do not apply exceptions to rules to your choice of answers to questions about the rules! Answer the question as it has been asked.

Here is an example of a question you might see:

1) *What is the maximum size of time-delay fuse that can be used for a 20hp 575V three-phase motor?*

    A) 30 A.   B) 35 A.   C) 40 A.   D) 45 A.

Rule 28-200 (a) of the CEC states that the motor shall be protected by an over-current device of a rating not to exceed the values in Table 29, using the full-load amps (FLA) of the motor. The FLA of the motor is 22 A. We can find the FLA by consulting Table 44 of the CEC. Next, we use Table 29, which states that a multiplier of 175% is to be used for time-delay fuses for squirrel cage (three-phase) motors. Therefore, the calculation is: 22 A × 1.75 = 38.5 A. This would be the maximum size, so the answer must be B: 35 A.

However, if you were referencing this section in a hurry, or you have read the full section before and don't check again to make sure you remember the rule correctly, you could apply an *exception* to the rule mentioned further down.

Rule 28-200 (d) (ii) states specifically that *if* the first over-current device will not allow the motor to start, the value may be increased to a maximum of 225% of the motor's full-load amps. Many apprentices might decide that the multiplier of 225% is the rule, and not an exception to it—simply because this is the one part of the section they remember during a high-pressure, timed exam. If so, they would use the following calculation: 22 A × 2.25 = 49.5 A. Since this gives the maximum size, the apprentice would come up with 45 A, and give the *incorrect* answer of D. Of course, Rule 28-200 (d) (ii) is an **exception** to the rule—and *only to be used if the first over-current device will not allow the motor to start.*

The answer to the question is B, 35 A.

Practice

Using the methods above, practise using all the tools in the code book to see how many different ways you can find the answer to a question! **And learn which tools you find easiest to use, or are best to use for certain things.**

Practice Questions: Using the CEC Book

Each question requires you to locate information from the CEC book. Try one or more approaches.

1) What is the maximum current carrying capacity of a neutral supported cable #2 NS90 aluminum with 3 insulated conductors?

2) What is the minimum length of free conductor that must be left in a box for making connections?

3) What is the minimum size of bonding conductor for an instrument transformer?

4) How is flat conductor cable (FCC) to be fastened to the floor?

5) What is the maximum voltage rating for a TW75 insulated conductor?

6) Can a panel board be located in a stairwell?

7) What is the maximum distance between a motor and a motor disconnecting means?

8) What is the maximum voltage to ground allowed in a dwelling unit?

9) What is the full-load current rating of a 2hp 240V DC motor?

10) What is the maximum current rating allowed before cables must be derated due to the effects of sheath currents?

Answers to Questions

In each answer, you can follow the method yourself to see how you get the answer, and which method worked better. With practice, your searches will become more efficient. The arrows in each explanation show logical paths, most consisting of only 2 or 3 steps, taking you from the question to the answer, sometimes using more than one method.

1) What is the maximum current carrying capacity of a neutral support cable #2 NS90 aluminum with 3 insulated conductors?

Using the Index: Begin with the main word followed by secondary words: *Cable, neutral-supported*. Result: Table 36.

Using the Index of Tables: Table 36A: Maximum Allowable Ampacity for Aluminum Conductor Neutral Supported Cables.

Note 1: Neutral supported cable has a bare neutral conductor that supports the insulated ungrounded or live conductors. Therefore, a cable with 3 insulated conductors will have 4 conductors in total and is referred to as "quadraplex."

Note 2: Although the *Index* cites Table 36 and the *index for the Tables* cites 36A, both are in fact Table 36A. When the latest CEC book was released, Table 36 was split into Table 36A and B, but the correction wasn't made to this entry in the Index.

Answer: Table 36A → 150 A.

2) *What is the minimum length of free conductor that must be left in a box for making connections?*

Using the Index: *Boxes and Fittings* (In a few cases, primary and secondary words are merged, as in this one) → Rule 12-3000.

Using the Table of Contents: Section 12: Wiring Method → Installation of Boxes → page 76.

Answer: Rule 12-3000 (5) → 150 mm.

3) *What is the minimum size of bonding conductor for an instrument transformer?*

Using the Table of Contents: Section 10: Grounding and Bonding → Grounding and Bonding Conductor → page 43.

Note: You identify that the most likely section to contain your answer will be Section 10, because *bonding* is the emphasis in the CEC book, and relatively small topics like *instrument transformer* will be handled elsewhere.

Answer: Rule 10-828 → conductor size #12.

4) *How is flat conductor cable to be fastened to the floor?*

Using the Table of Contents: Section 12: Wiring Methods → Conductors → Flat Conductor Cable → page 59.

Answer: Rule 12-818 → With adhesive.

5) What is the maximum voltage rating for a TW75 insulated conductor?

Using the Index: Begin with main and secondary words: *Voltage, rating of conductors and cables*. Result: Table D1.

Using the Index of Appendix D: Table D1.

Answer: Table D1 → 600 V.

6) Can a panel board be located in a stairwell?

Using the Index: Begin with main and secondary words: *Panel boards, location of*. Result: Rule 26-402.

Using the Table of Contents: Section 26: Installation of Electrical Equipment → Panel Boards → page 144.

Answer: Rule 26-402 (1) → No.

7) What is the maximum distance between a motor and a motor disconnecting means?

Using the Index: Begin with main and secondary words: *Motor, disconnecting means*. Result: Rule 28-604.

Using the Table of Contents: Section 28: Motors and Generators → Disconnecting Means → page 161.

Answer: Rule 28-604 (3) (a) → Within sight and within 9 m.

8) What is the maximum voltage to ground allowed in a dwelling unit?

Using the Index: Begin with main and secondary words: *Voltage, in dwelling units*. Result: Rule 2-106.

Answer: Rule 2-106 → 150 V.

9) What is the full-load current rating of a 2hp 240V DC motor?

Using the Index: Begin with main and secondary words: *Motor, currents*. Result: Tables 44, 45, and D2.

28

Using the Index of Appendix D: Table D2.

Answer: Table D2 → 8.5 A.

10) What is the maximum current rating allowed before cables must be derated due to the effects of sheath currents?

Using the Index: Begin with main and secondary words: *Sheath currents,* Rule 4-008. (This is another case of main and secondary words being merged.)

See Appendix B: Rule 4-008, the paragraphs under the heading *Single Conductors in Free Air*.

Answer: Rule 4-008 (see Appendix B) → 425 A.

Practice Exam 1 5

The following practice exam has 100 questions. The answer key can be found on page ... and the explanations for the answers begin on page 71.

Occupational Skills

1) During a construction project, there is a conflict over the location of a heat register and a light fixture. Which should be consulted?

A) Facility engineer. B) Duct work manufacturer.
C) Project owner. D) General contractor.

2) If a worker refuses to work because they feel working conditions are unsafe, what is the proper course of action?

A) Worker goes home and waits to be notified of a resolution.
B) Ministry of Labour representative is brought in to evaluate the working conditions.
C) Employer or supervisor investigates together with on-site health and safety representative.
D) Worker's concerns are addressed and work resumes.

3) When lifting an object with a sling and a crane, what sling angle will result in the least amount of stress on the sling?

A) 45°. B) 60°. C) 75°. D) 90°.

4) When replacing the blade on a hacksaw, the teeth on the blade should be

   A) Pointing toward the handle.
   B) Pointing away from the handle.
   C) Straight.
   D) The blade only fits one way.

5) A 9 m extension ladder is extended 6.83 m up the side of a wall. How far should the base of the ladder be from the wall?

   A) 1.6 m.   B) 1.9 m.   C) 2.5 m.   D) 2.8 m.

6) When testing a high-voltage cable, you should use a

   A) High potential tester.   B) Wheatstone bridge.
   C) Clamp-on ammeter.   D) Watt meter.

7) If a millwright installs a lock and tag on a piece of equipment, who is allowed to remove the lock and tag?

   A) A supervisor.
   B) The millwright.
   C) The foreman starting the next shift.
   D) Anyone the management has given a key to.

8) When new materials are received on a job site, what is to be done with the material safety data sheet (MSDS)?

   A) Read and discard.
   B) Tape it to the container(s).
   C) Place it in the on-site MSDS file folder.
   D) Give it to the supervisor.

9) Anyone who is to perform work in a certified trade must file their contract of apprenticeship

   A) Within 2 years.
   B) Within 1 year.
   C) Within 3 months.
   D) Immediately.

10) Which code or standard determines whether a particular building requires a fire alarm system?

    A) The National Building Code (NBC).
    B) The Canadian Electrical Code.
    C) The National Fire Code.
    D) The Underwriters' Laboratories of Canada Standard for the Installation of Fire Alarm Systems to Certification.

11) When the complexity of electrical work makes it difficult to determine exactly how long the work will take to complete, the appropriate method of payment to use is

    A) Lump sum.
    B) Unit price.
    C) Time and material.
    D) Paid when complete.

12) In what type of drawing would you find the depth of an open web steel joist?

    A) Shop drawing.   B) Structural drawing.
    C) Electrical drawing.   D) Mechanical drawing.

Block B: Distribution and Services

13) A three-phase three-wire system has ground-fault indicator lights blinking. What will happen if A-phase goes to ground?

    A) A-phase light goes on.   B) A-phase light goes off.
    C) All lights go on.   D) All lights go off.

14) What is the minimum size of single-conductor aluminum-sheathed cable that can be used for a consumer's service?

    A) #6 AWG.   B) #5 AWG.   C) #4 AWG.   D) #3 AWG.

15) What is the minimum head room allowed by the code when working around a motor control centre where bare live parts are exposed?

    A) 1 m.   B) 1.5 m.   C) 2 m.   D) 2.2 m.

16) We are permitted to replace two-prong receptacles with three-prong receptacles, provided we

   A) Bond the three-prong receptacle to the outlet box.
   B) Protect the branch circuit with an arc-fault circuit interrupter (AFCI).
   C) Protect the branch circuit with a Class-A ground-fault circuit interrupter (GFCI).
   D) Install a GFCI receptacle into the branch circuit.

17) Where single conductors enter metal boxes through separate openings, precautions must be taken to prevent overheating of the metal by induction if the current carried per conductor exceeds how many amps?

   A) 200 A.   B) 250 A.   C) 400 A.   D) 600 A.

18) When installing the grounding system for an outdoor high-voltage station, the minimum size of ground grid conductor must be

   A) #3/0 AWG.   B) #2/0 AWG.   C) #3 AWG.   D) #6 AWG.

19) When installing a backup power source, the type of device used to switch from normal power to backup power is a

   A) Three-phase switch.
   B) Single-pole double-throw switch.
   C) Single-pole switch.
   D) Transfer switch.

20) What is the maximum voltage drop permitted for a 600V three-phase four-wire feeder in an industrial plant?

   A) 18 V.   B) 25 V.   C) 30 V.   D) 60 V.

21) One of the main components of all uninterruptible power supply (UPS) units is the

   A) Auto transformer.   B) Differential relay.   C) Battery.   D) Capacitor.

22) What is the minimum size of copper TW75 conductor in conduit required to supply a 50 kVA (kilo-volt amps) three-phase dry type transformer, with a 600V primary and a 120V/208V secondary?

   A) #2 AWG.   B) #3 AWG.   C) #4 AWG.   D) #6 AWG.

23) Which of the following types of flexible cords is suitable to supply a temporary panel?

   A) SJ.   B) SOW.   C) STO.   D) SJOW.

24) The 2 main reasons to provide electrical systems with surge protection are lightning and

   A)   Utility switching.   B)   Poor power factor.
   C)   Harmonic distortion.   D)   DC generators.

25) The minimum size of Teck90 multi-conductor cable that can be used to supply a 60A 120V/240V panel in an apartment unit is

   A) 6/3.   B) 4/3.   C) 3/3.   D) 2/3.

26) What is the resulting kilo-volt amps of three 100kVA transformers connected delta if one of the transformers fails and the 2 remaining transformers are connected open delta?

   A) 300 kVA.   B) 225 kVA.   C) 173 kVA.   D) 150 kVA.

27) A Type 4 enclosure is designed for use where subject to

   A) Class 1 location.   B) Oil spray.   C) Submersion.   D) Splashing water.

28) What is the maximum size of non-time delay fuse that can be used to protect a 600V three-phase 25kVAr (kilo-volt amps reactive) capacitor?

   A) 70A.   B) 60A.   C) 40A.   D) 30A.

29) The minimum size of grounding conductor for a service using conductors with a 200A rating is

   A) #8 AWG.   B) #6 AWG.   C) #4 AWG.   D) #3 AWG.

30) The minimum bending radii for a 45mm diameter 5kV portable power cable is

   A) 270 mm.   B) 315 mm.   C) 360 mm.   D) 540 mm.

31) What is the maximum unguyed projection for the support member, or service mast, for a consumer's service?

   A) 750 mm.   B) 1 m.   C) 1.5 m.   D) 2.5 m.

32) Stress cones, when used to terminate high-voltage cables, are designed to

   A) Relieve the stress on the connection due to the weight of the cable.
   B) Allow the cable to fit the connector more easily.
   C) Drain induced static voltage to ground.
   D) Stress cones are not used with high-voltage cables.

33) Power factor correction capacitors are installed on electrical systems to raise the power factor above

   A) Unity.   B) 98%.   C) 95%.   D) 90%.

34) Prior to opening the switch on the primary of a large transformer, it is a good practice to

   A) Open the secondary loads.
   B) Open the switch as quickly as possible to reduce any arcing.
   C) Open the primary fuses.
   D) Disconnect and ground the feeders.

35) Metering equipment shall be allowed to be connected to the supply side of the service box when

   A) No live parts or wiring are exposed.
   B) The supply side is AC and the voltage exceeds 300 V between conductors.
   C) The service exceeds 200 A.
   D) The meter is mounted on a 200A meter socket.

36) When crossing over a driveway to a residential garage, supply conductors to a consumer's service must maintain a minimum height above grade of

   A) 3 m.   B) 3.5 m.   C) 4 m.   D) 5.5 m.

Block C: Branch Circuit Wiring

37) What is the minimum coverage for a 120V/208V three-phase service using TWU 75 conductors that are direct buried in the earth in an area that is not subject to vehicular traffic?

   A) 1 m.   B) 900 mm.   C) 750 mm.   D) 600 mm.

38) When connecting a portable motor, which of the following wiring methods should be used?

   A) SO.   B) TEW.   C) MND 90.   D) RA-90.

39) A heating cable set is to be buried in concrete to melt the snow that may accumulate on a walkway leading into a building. What is the minimum depth of the cable set if the area is not subject to vehicular traffic?

   A) 50 mm.   B) 100 mm.   C) 150 mm.   D) 200 mm.

40) For a luminaire to be controlled at 4 locations, what combination of switches is required?

   A) 2 three-way switches and 2 single-pole switches.
   B) 2 four-way switches and 2 single-pole switches.
   C) 2 three-way switches and 2 four-way switches.
   D) 2 three-way switches, 1 four-way switch, and 1 single-pole switch.

41) When an AC-90 cable is run through metal studs that are to be covered by drywall, how far from the edges of the studs must the cable be kept?

A) 6 mm. B) 19 mm. C) 24 mm. D) 32 mm.

42) What is the maximum voltage allowed on an NMD-90 cable between any two conductors?

A) 150 V. B) 240 V. C) 300 V. D) 600 V.

43) What is the expansion of 40 m of a PVC conduit in a temperature range of −22°C to 33°C?

A) 156.4 mm. B) 114.4 mm. C) 68.4 mm. D) 45 mm.

44) What is the number of #14 AWG conductors that can be installed in a 3" × 2" × 3" device box that has a single-pole switch and three-wire connectors?

A) 4. B) 5. C) 6. D) 7.

45) What is the maximum number of #10 R90 (600V unjacketed) power conductors that can be installed in a 75mm × 75mm wireway?

A) 65. B) 70. C) 104. D) 200.

46) Which of the following lighting sources requires an igniter?

A) High-pressure sodium. B) Mercury vapour.
C) Instant-start fluorescent. D) Incandescent.

47) What is the distance that the heating portion of a heating panel must be kept from an outlet box that is being used as a luminaire?

A) 600 mm. B) 200 mm. C) 50 mm. D) 13 mm.

48) When four #3 R90 and four #4 R90 (both XLPE 600V without jacket) are to be pulled into 1 conduit, the minimum size of raceway is

   A) 27 mm.   B) 35 mm.   C) 41 mm.   D) 53 mm.

49) When installing a light post to a concrete base with 4 bolts, and the post is not sitting level on the base, you should

   A)   Remove the post and level the concrete.
   B)   Level the lower side with a steel shim.
   C)   Drill and tap new bolts into the base.
   D)   Send the post back to the manufacturer.

50) Heat pumps used in residential applications can be used to provide

   A)   Heating or cooling.
   B)   Heating only.
   C)   Cooling only.
   D)   Heat pumps are not used in residential applications.

51) If 16mm and 53mm ridged metal conduit are supported on the same hangers, what is the maximum distance between the hangers?

   A) 3 m.   B) 2.5 m.   C) 2 m.   D) 1.5 m.

52) Which of the following is recommended as an emergency light source?

   A)   Incandescent.   B)   Metal halide.
   C)   High-pressure sodium (HPS).   D)   Mercury vapour.

53) What is the maximum distance from the floor to the midpoint of a receptacle used to supply an electric range in a dwelling unit?

   A) 110 mm.   B) 120 mm.   C) 130 mm.   D) 140 mm.

54) When installing a lightning arrester you should

   A) Place the device outdoors.
   B) Keep the conductor as short as possible.
   C) Only use if the building does not have lightning rods.
   D) Only use if the building does have lightning rods.

55) Storage batteries used for emergency power shall maintain what percentage of their voltage rating for the amount of time required by the National Building Code of Canada?

   A) 100%.   B) 95%.   C) 91%.   D) 85%.

56) When installing a device box in a concrete wall, what is the maximum distance the box may be recessed?

   A) 2 mm.   B) 4 mm.   C) 6 mm.   D) The box cannot be recessed.

57) The correction factor for an R90 conductor used in an ambient temperature of 50°C is

   A) 0.81.   B) 0.8.   C) 0.74.   D) 0.75.

58) The minimum size of copper conductor for a 6V emergency lighting system that has 4 A of current and must be 15 m in length is

   A) #14 AWG.   B) #12 AWG.   C) #10 AWG.   D) #8 AWG.

59) In a Class I, Zone 1 area, what is the maximum distance that a seal may be placed from an enclosure that contains an arcing device?

   A) 50 mm.   B) 100 mm.   C) 300 mm.   D) 450 mm.

60) When determining the length of a pull box that has 250kcmill conductors in a ridged metal conduit and the pull is a straight pull, the length must be at least how many times the size of the largest raceway?

   A) 4 times.   B) 6 times.   C) 8 times.   D) 10 times.

Block D: Motor and Control Systems

61) By inserting an external resistance into the shunt field of a compound DC motor, the result will be

   A) The motor speed will reduce.
   B) The motor speed will increase.
   C) The motor speed will remain unchanged.
   D) This should never be attempted.

62) When using a wye-delta motor starter, the motor will receive approximately what value of full-line voltage upon start-up?

   A) 58%.   B) 77%.   C) 86%.   D) 100%.

63) How do you reverse the direction of a three-phase AC motor?

   A) Interchange the start and run winding.
   B) Direction cannot be changed.
   C) Interchange any 2 supply leads.
   D) Turn the motor around.

64) Instantaneous trip circuit breaker used with motor-starting applications has been designed to provide

   A) Overload protection.   B) Short-circuit protection.
   C) Overheating protection.   D) All of the above.

65) When smooth control over the speed of an AC motor is required, the best way to achieve this is to use a

   A) Rheostat.   B) Timing relay.
   C) Variable frequency drive.   D) Resistor bank.

66) The correct action to take if a motor trips on thermal overload is to

   A) Allow a cooling period, then reset.
   B) Reset the overload immediately.
   C) Install new overloads.
   D) Change the fuses.

67) A start-stop station is run in a ridged metal conduit to remotely control a motor. The starter has a coil and a holding contact. How many wires must be brought to the remote station?

   A) 2.   B) 3.   C) 4.   D) 6.

68) A 20 horsepower 575V three-phase refrigeration compressor is unable to start with the proper size of time-delay fuse. What is the maximum size that the fuse can be increased to?

   A) 35A.   B) 45A.   C) 60A.   D) 80A.

69) When selecting an over-current device for motor-starting applications, the one that can be set the lowest, and therefore allow the lowest level of fault current through, is

   A) Type P.   B) Type HRC.   C) Type R.   D) Type J.

70) What is the minimum size of R90 copper conductor in conduit that is required to supply a 575V three-phase squirrel cage motor with an insulation class of B and a full-load amps of 82 A?

   A) #1 AWG.   B) #2 AWG.   C) #3 AWG.   D) #4 AWG.

71) What is the minimum allowable size of R90 copper conductor for use to supply the secondary resistors of a 575V three-phase 50hp Class-B insulation rating wound-rotor motor?

   A) #10 AWG.   B) #8 AWG.   C) #6 AWG.   D) #4 AWG.

72) If a motor states the maximum over-current device is 30 A, and the Canadian Electrical Code rules state it could be 40 A, which over-current device should be used?

   A) 45.   B) 40.   C) 35.   D) 30.

73) A 240V single-phase motor is to be protected by an overload relay. How many overload devices are required?

   A) 3.
   B) 2.
   C) 1.
   D) Single-phase motors do not require overload protection.

74) The disconnecting means for refrigeration equipment shall be located within sight and _____ metres away.

   A) 3.   B) 5.   C) 7.   D) 9.

75) When a motor is controlled by a float switch, this is an example of what type of control?

   A) Two-wire control.   B) Three-wire control.
   C) Low-voltage protection.   D) Manual control.

76) Two or more motors shall be permitted to be grouped under the protection of a single set of over-current devices, provided that the ampacity of the over-current device does not exceed

   A) 150% of the largest FLA.
   B) 60 A.
   C) 30 A.
   D) 15 A.

77) When a motor is placed in an ambient temperature of 40°C, the motor supply conductor insulation temperature rating must be

   A) Corrected using Table 5A of the CEC.
   B) Increased by 10°C.
   C) Derated using Table 5C.
   D) Nothing need be done if the motor is rated for 40°C.

78) What is the minimum size of R90 (XLPE) conductor in conduit that can be used to connect a 60hp 3ph 575V squirrel cage motor?

   A) #6 AWG.   B) #4 AWG.   C) #3 AWG.   D) #2 AWG.

79) The speed of a three-phase squirrel cage motor depends primarily on

　　A)　Line voltage and over-current device size.
　　B)　Line current and the horsepower of the motor.
　　C)　Brush position and spring tension.
　　D)　System frequency and the number of poles.

Block E: Extra Low Voltage Systems

80) If the value of the end-of-line resistor is 3,900 Ω and the supply voltage is 24 V, the supervisory current is

　　A) 0.06 A.　　　B) 0.006 A.　　　C) 0.0006 A.　　　D) 0.00006 A.

81) Coaxial cable shall be permitted to be used for connection between the cable distribution plant and the customer service enclosure provided that

　　A)　For a single dwelling the open circuit voltage does not exceed 90 V.
　　B)　For a single dwelling the open circuit voltage does not exceed 70 V.
　　C)　For a single dwelling the open circuit voltage does not exceed 50 V.
　　D)　For a single dwelling the open circuit voltage does not exceed 30 V.

82) Conductive optical fibre cables shall be permitted in the same raceway with

　　A)　Conductors with the same insulation rating.
　　B)　Conductors 30 V or less.
　　C)　Class 2 conductors.
　　D)　Power and lighting conductors.

83) Which of the following diagrams is correct for a fire alarm signal circuit?

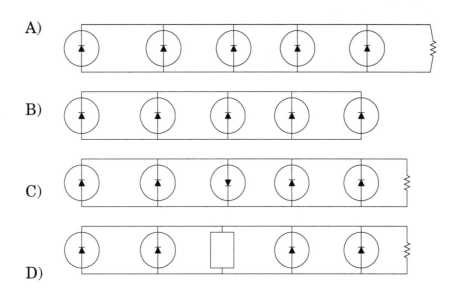

84) When installing magnetic contacts on multiple windows in a single zone of a security system, the connection in the panel for the window contacts must be made

A) In combination. B) In code. C) In parallel. D) In series.

85) What is the smallest size of conductor that can be pulled into a raceway for use in a fire alarm system?

A) #14 AWG. B) #16 AWG. C) #18 AWG. D) #22 AWG.

86) The use of communication flat cable (CFC) shall be permitted to be used

A) In wet locations. B) In dwelling units.
C) Under carpet squares. D) In patient care areas.

87) Computer systems must have hardware and software to operate properly. An example of software would be

A) An optical mouse. B) A server. C) A program. D) A printer.

88) A two-stage fire alarm system will give authorized personnel how much time to investigate the possibility of a fire before proceeding to evacuation mode when an initiating device has been activated?

   A) 1/2 hour.   B) 15 minutes.   C) 8 minutes.   D) 5 minutes.

89) An emergency pull cord for a nurse call system is usually located in

   A) Washrooms.   B) Toilets.   C) Showers.   D) All of the above.

Block F: Upgrading, Maintenance, and Repair

90) If due to a fault, 1 phase of a 600V three-phase three-wire system was grounded, the voltage of either of the remaining 2 phases to ground should be

   A) 600 V.   B) 347 V.   C) 300 V.   D) 208 V.

91) What do you use or do to prevent galvanic corrosion between aluminum and copper conductors?

   A) Aluminum joint compound.   B) Cover the connection with grease.
   C) Keep the connection dry.   D) Use split bolt connectors.

92) Calculate the total resistance of the following circuit

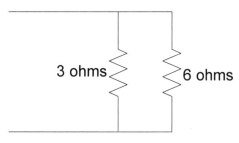

   A) 9Ω.   B) 5Ω.   C) 3Ω.   D) 2Ω.

93) A 4-20 mA signal transmitter is being used to monitor 0–1,200 V. What would the voltage measurement be if the transmitter was putting out 7 mA?

   A) 525 V.   B) 450 V.   C) 225 V.   D) 75 V.

94) You are called to troubleshoot a piece of equipment that has been in service for some time. The equipment is controlled by a programmable logic controller (PLC). The program is running on the PLC, but one rung will not come true. Why?

A) Output mismatch.
B) Input is not functioning.
C) Counter is done.
D) Computer processing unit (CPU) not working.

95) When conducting a thermo photography inspection of a building's electrical system, you should

A) Open all panel doors prior to test to allow them to cool.
B) Open all loads before test.
C) Make sure all loads remain connected during test.
D) Turn off all lighting loads to prevent false readings.

96) In the diagram below, which of the fuses has been blown?

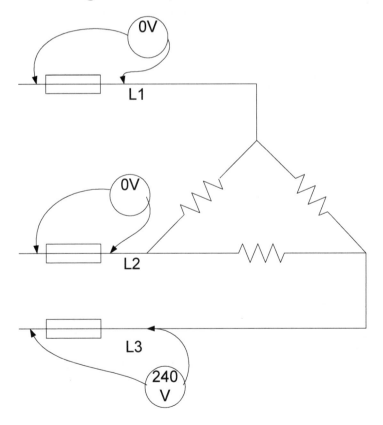

A) Fuse L1.   B) Fuse L2.   C) Fuse L3.   D) Fuse L1 and L2.

97) A resistor has a colour-coding of red, blue, orange and gold. What is the value of the resistor?

　　A)　2,600Ω, 5% tolerance.
　　B)　26,000Ω, 5% tolerance.
　　C)　26,000Ω, 10% tolerance.
　　D)　2,600Ω, 10% tolerance.

98) When not in use, the state of the charge of a lead-acid battery should be

　　A)　Fully charged.
　　B)　Charged once a year.
　　C)　About 1/2 charge.
　　D)　Let the battery fully discharge, then recharge.

99) What is the reason we install all 4 conductors of a three-phase four-wire system in one metal conduit?

　　A)　Easier to install.　　　　　　B)　Easier to troubleshoot.
　　C)　Reduces cost.　　　　　　　　D)　Reduces induction.

100) When the speed of an AC motor needs to be controlled and the existing method is a wound rotor motor, you would upgrade the control by installing a squirrel cage motor and a

　　A)　UPS.
　　B)　Variable frequency drive (VFD).
　　C)　Field relay.
　　D)　Drum switch.

Practice Exam 2   6

The following practice exam has 100 questions. The answer key can be found on page 70 and the explanations for the answers begin on page 91.

Block A: Occupational Skills

1) When an electrical contractor is to carry out electrical work, an inspection permit shall be obtained from the inspection department

   A) At some point during the work.
   B) Before the walls are covered.
   C) Before the work starts.
   D) Within the same calendar year.

2) The responsibility of keeping track of apprentices' hours of work for the purpose of completing an apprenticeship is the responsibility of

   A) The employer.
   B) The apprentice.
   C) The ministry.
   D) Only competencies must be recorded, not hours.

3) The lock out–tag out procedures found in the Occupational Health and Safety Act (OHSA)

   A) Are the minimum standards, but individual companies may raise the standards.
   B) Are the only accepted procedures and must be followed exactly.
   C) Are to be used only as a guide for the safety representative.
   D) Need only be applied when non-electrical staff is involved in the repairs.

4) Which of the following are used to indicate the necessary level of flame spread rating in a building of combustible construction?

   A) FT1.   B) FT2.   C) FT4.   D) FT6.

5) How old must a person be to become an electrical apprentice?

   A) 14.   B) 15.   C) 16.   D) 18.

6) What type of drawing would show you all the major components of a building's electrical system?

   A) Shop drawing.   B) Single-line diagram.
   C) Plan view.   D) Elevation.

7) What is the meaning of the following WHMIS symbol?

   A) Flammable and combustible materials.
   B) Materials that are safe for handling.
   C) Materials that could cause a biohazard.
   D) Materials causing immediate and serious toxic effects.

8) A drawing has a scale of 1:20 mm. If a line on the drawing is 75 mm in length, what length does the line represent?

A) 0.75 m. B) 1.25 m. C) 1.5 m. D) 1.75 m.

9) The hand signal used to instruct the crane operator to stop is

A) Forefinger pointing up, hand moving in small circles.
B) Arm extended palm down move hand right and left.
C) Arm extended forward hand open at right angle pointing upward.
D) Arm extended downward hand making a fist.

10) When performing a test on a live circuit downstream from the service entrance equipment, your meter must have a Category _____ rating?

A) 4. B) 3. C) 2. D) 1.

11) When working on a construction site and painting and trim work is being done by other tradespeople, which of the following electrical activities is most likely being done?

A) Embedded work.
B) Lighting fixtures.
C) Trenching.
D) Branch circuits and pulling wire.

12) Waste material and debris should be removed from work and access areas at least

A) Once a day.
B) Before each work break.
C) Immediately.
D) Before the job is complete.

Block B: Distribution and Services

13) A residential building has an electric heating load of 20 kW. What part of this load must be added to the service calculation?

   A) 20 kW.   B) 17.5 kW.   C) 15 kW.   D) 50% of load.

14) What is the minimum size of neutral support cable (NS75 aluminum) that can be used to supply a 100A 208V three-phase four-wire service?

   A) #1.   B) #2.   C) #3.   D) #4.

15) What is the minimum size of conduit for a 100A 120V/240V single-phase service using R90 unjacketed conductors?

   A) 16 mm.   B) 2 mm.   C) 27 mm.   D) 35 mm.

16) What is the minimum size RWU 90 conductor that is required for a 1000A 347V/600V three-phase service, using the IEEE installation configuration detail 3 of B4-3?

   A) 350 kcmill.   B) 500 kcmill.   C) 750 kcmill.   D) 1,000 kcmill.

17) What is the calculated wattage for the following single-family dwelling?

House Area	127 m²
Electric Range	12 kW
Water Heater	4 kW
Pool Heater	4 kW
Air Conditioning	5 kW
Building system	120V/240V single-phase

   A) 20 kW.   B) 21 kW.   C) 21.5 kW.   D) 22 kW.

18) The minimum number of ground rods required for grounding an outdoor high-voltage pad-mount transformer is

A) 2.　　　　B) 3.　　　　C) 4.　　　　D) 6.

19) You are installing a 600A feeder using single-conductor cables. In order to eliminate sheath currents, you should

A)　Use cable connectors that can accommodate bonding jumpers.
B)　Use non-ferrous metallic plates at both ends of the conductors.
C)　Ground the supply side and isolate the load side.
D)　Isolate both the supply and the load side from ground.

20) The potential difference between a grounded metal structure and a point on the earth's surface separated by a distance equal to the normal horizontal reach is referred to as

A)　The step voltage.　　　　B)　The touch voltage.
C)　The separation voltage.　　D)　The potential voltage.

21) An uninterruptible power supply uses what device to supply the load with alternating current from the battery?

A) Inverter.　　B) Transformer.　　C) Coil.　　D) Capacitor.

22) Unless special permission is granted, a 120V/240V supply service may supply the following number of consumer services in a building:

A) 6.　　　　B) 4.　　　　C) 2.　　　　D) 1.

23) What is the maximum spacing between fence posts used for the guarding of outdoor electrical equipment?

A) 1 m.　　　B) 3 m.　　　C) 4.5 m.　　　D) 6 m.

24) What is the approximate available fault current on the secondary of a 150 kVA transformer that has a 600V three-phase primary and a 120V/208V three-phase secondary with an impedance of 5%?

A) 8,300 A. B) 10,200 A. C) 14,300 A. D) 23,000 A.

25) When installing a wiring system that uses an ungrounded supply, you must provide

A) Ground-fault protection.
B) A ground grid.
C) Ground detection device.
D) You cannot have an ungrounded supply.

26) When determining the living area of a dwelling unit for a service calculation, you must consider 100% of the ground floor, 100% of the living area of the second floor, and _____ % of the basement floor.

A) 30%. B) 50%. C) 75%. D) 100%.

27) The service head location of a consumer's service must be in a location

A) That is in compliance with the supply authority.
B) That is accessible to authorized personnel.
C) That is easily accessible.
D) That cannot be reached from the building rooftop.

28) What is the calculated wattage of the basic load for the following storage warehouse when determining the service

The area of the warehouse is 35 m × 20 m (outside dimensions).

A) 10,000 W.   B) 7,000 W.
C) 3,500 W.   D) 2,450 W.

29) Ground-fault protection shall be provided to de-energize all normally ungrounded conductors under which set of conditions?

A) In solidly grounded systems rated more than 150 V to ground, and less than 750 V phase to phase and 1000 A or more.
B) In ungrounded systems rated more than 150 V to ground, and less than 750 V phase to phase and 1000 A or more.
C) In solidly grounded systems rated less than 150 V to ground, and less than 600 V phase to phase and 1000 A or more.
D) In ungrounded systems rated more than 150 V to ground, and less than 600 V phase to phase and 1000 A or more.

30) A moulded case circuit breaker shall be permitted to be installed in a circuit having an available fault current higher than its rating provided that

A) The circuit breaker is installed in an explosion-proof enclosure.
B) The circuit breaker is installed in a NEMA Type 4 enclosure.
C) The circuit breaker is a recognized component of a series-rated combination.
D) The circuit breaker has ground-fault circuit protection.

31) What is the minimum number of branch circuit positions required for a 100A panel board used in a dwelling unit?

A) 24. B) 30. C) 40. D) 48.

32) When the method of grounding a service is a plate electrode, the ground plate must be in direct contact with the exterior soil and buried at least

A) 1,200 mm below grade.   B) 1,000 mm below grade.
C) 600 mm below grade.   D) 300 mm below grade.

33) What is the calculated load on the 120V/240V single-phase service equipment in amps for the following two-unit apartment?

	Unit A	Unit B
Area	47 m²	43 m²
Electric range	11 kW	10 kW
Electric heat	8 kW	7 kW

A) 147.9 A.   B) 128.8 A.   C) 115.5 A.   D) 100 A.

34) The minimum size of grounding conductor for a service with conductors rated at 1000 A is

A) 250 kcmill.   B) 4/0 AWG.   C) 3/0 AWG.   D) 2/0 AWG.

35) Some distribution transformers are equipped with tap changers. The purpose of the tap changer is to

A) Open the primary circuit to the transformer.
B) Close the secondary circuit to the transformer.
C) Change the primary ratio of the transformer to the secondary ratio of the transformer.
D) Obtain slightly different voltage ratios.

36) When connecting a lightning arrester to the line conductor, the connection shall be

A) Made with an insulated conductor.
B) As short and as straight as possible.
C) Made with a thermit weld.
D) Outdoor on the incoming pole.

Block C: Branch Circuit Wiring

37) A conduit run contains four #6 R90 (XLPE) (without jacket) current carrying conductors and has an ambient temperature of 50°C. What is the maximum current carrying capacity of any one of the conductors?

A) 65 A.   B) 52 A.   C) 41 A.   D) 38 A.

38) An electrical junction box has an area of 775 ml stamped into the metal. What is the maximum number of #10 AWG conductors that may be installed in the box?

A) 21.   B) 22.   C) 23.   D) 24.

39) A run of ridged metal conduit contains twenty #12 T-90 nylon conductors. What is the minimum size of conduit that can be used?

A) 27 mm.   B) 35 mm.   C) 53 mm.   D) 63 mm.

40) A bare or un-insulated bonding conductor may be run in a conduit that contains ungrounded conductors of the same circuit provided that the run

    A) Not exceed 30 m and 4 quarter bends.
    B) Not exceed 30 m and 2 quarter bends.
    C) Not exceed 15 m and 4 quarter bends.
    D) Not exceed 15 m and 2 quarter bends.

41) What is the minimum distance that a ground-fault circuit interrupter receptacle must be from the water's edge of a permanently installed swimming pool?

    A) 1.5 m.   B) 3 m.   C) 5 m.   D) 7.5 m.

42) If an electrical circuit has the voltage doubled and the resistance is reduced by half, what will happen to the current in the circuit?

    A) The current will decrease by half.
    B) The current will increase by 2 times.
    C) The current will increase by 4 times.
    D) The current will stay the same.

43) Which is the correct wiring diagram for a receptacle that has one half switched and one half with constant power?

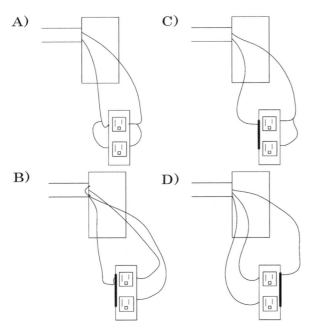

44) If the phase current of a delta system is 100 A, what is the line current?

   A) 57.7 A.   B) 86.6 A.   C) 100 A.   D) 173 A.

45) Ridged PVC conduit shall not be used in an area having an ambient temperature greater than

   A) 30°C.   B) 45°C.   C) 60°C.   D) 75°C.

46) Which of the following lighting sources provides the highest degree of efficacy?

   A) Low-pressure sodium.
   B) Mercury vapour.
   C) Instant-start fluorescent.
   D) Incandescent.

47) Where the heating portion of a heating cable set is not totally embedded in non-combustible material, the rating—or setting—of the branch circuit over-current device shall not exceed

   A) 15 A.   B) 20 A.   C) 30 A.   D) 60 A.

48) What is the designation for a 30A 250V receptacle used for a residential clothes dryer?

   A) 5-15R.   B) 14-30P.   C) 14-30R.   D) 14-50R.

49) What is the maximum ampacity rating for a #8 AWG Type-S flexible cord that has 4 current carrying conductors?

   A) 45 A.   B) 35 A.   C) 28 A.   D) 25 A.

50) What is the maximum setting of the high-limit protection for an electric storage-tank hot water heater?

   A) 96°C.   B) 90°C.   C) 70°C.   D) 55°C.

51) What is the area classification of the nozzle boot of a gasoline dispenser?

   A) Class I Zone 3.
   B) Class I Zone 2.
   C) Class I Zone 1.
   D) Class I Zone 0.

52) Which of the following light sources does not use a ballast?

   A) Metal halide.   B) Fluorescent.   C) Neon.   D) High-pressure sodium.

53) What is the minimum thickness of a running board used for the protection of wires and cables?

   A) 12 mm.   B) 19 mm.   C) 25.4 mm.   D) 40 mm.

54) When installing a run of electrical metallic tubing (EMT), what is the maximum distance that may be run from a box before a strap is required?

   A) 300 mm.   B) 600 mm.   C) 1 m.   D) 1.5 m.

55) What type of machine screw is used to connect a keyless lamp holder to a 4" × 1½" octagon box?

   A) 6/32.   B) 8/32.   C) 10/32.   D) 10/24.

56) A junction box has a 53 mm conduit entering on one side and another 53 mm conduit exiting on the opposite side. The 2 conduits are being used for a straight pull. What is the minimum distance between the 2 conduits?

   A) 53 mm.   B) 106 mm.   C) 318 mm.   D) 424 mm.

57) When installing a panel board in a dwelling unit, it shall be as high as possible, with an over-current device positioned higher than _____ metres above the floor.

   A) 1.7.   B) 2.   C) 2.1.   D) 2.5.

58) What is the maximum voltage between any 2 conductors of a non-metallic sheathed cable?

   A) 150 V. B) 250 V. C) 300 V. D) 600 V.

59) What is the maximum allowable conductor temperature in Celsius for an extra low voltage control (ELC) cable type?

   A) 150°C. B) 100°C. C) 90°C. D) 60°C.

60) A receptacle has an orange triangle on its face. Where is this type of receptacle used?

   A) In isolated ground installations.
   B) In hospitals.
   C) In schools.
   D) In airport hangars.

Block D: Motor and Control Systems

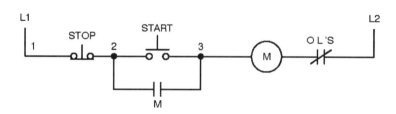

61) Referring to the figure above, the normally open M contact is generally referred to as a

   A) Maintaining contact. B) Sealing contact.
   C) Holding contact. D) All of the above.

62) What type of DC motor should never be connected to a belt drive system?

   A) Compound. B) Shunt.
   C) Series. D) Universal.

63) If an alternating current motor rated for continuous duty service has a marked service factor of 1.2, the multiplier used to determine the maximum rating of the overload device is

    A) 1.　　　　B) 1.15.　　　　C) 1.2.　　　　D) 1.25.

64) The approximate full-load amps of a 20hp 575V three-phase squirrel cage motor would be

    A) 10 A.　　　　B) 20 A.　　　　C) 30 A.　　　　D) 40 A.

65) The term used to describe a control circuit that operates on the failure of voltage to cause and maintain the interruption of power to the main circuit is called

    A)   Low-voltage protection.　　　　B)   Low-voltage release.
    C)   Two-wire control.　　　　　　　D)   Category 2 control.

66) If the centrifugal switch on a single-phase motor is left in the OPEN position on start-up due to a broken spring, the likely result will be

    A)   The run winding will burn out.
    B)   The start winding will burn out.
    C)   The motor will run at 50% of rated speed.
    D)   The motor may not start.

67) When an auto transformer is used as part of a motor control circuit, the purpose of the auto transformer is to

    A)   Increase the voltage on start-up.
    B)   Decrease the voltage on start-up.
    C)   Change the connection from delta to a wye.
    D)   Help slow the motor.

68) To change the direction of a DC motor, it is most common to

    A)   Change the shunt field connection.
    B)   Change the series field connection.
    C)   Change the armature connection.
    D)   Change the commutator field connection.

69) The centrifugal switch of a single-phase alternating current motor will open and disconnect the start winding at approximately what percent of rated motor speed?

A) 100%. B) 75%. C) 50%. D) 25%.

70) If a DC motor is supplied from a full-wave bridge rectifier and one of the diodes becomes defective, the most likely result to the motor will be

A) The motor will reverse direction.
B) The motor will stop.
C) The motor will slow.
D) The motor will work normally until a second diode fails.

71) Inverse-time circuit breakers used to protect a branch circuit for a one-phase motor shall be set or rated for a maximum of what percentage of the motor's full-load amps?

A) 80%. B) 125%. C) 175%. D) 250%.

72) "Jogging" a motor is a term used to describe

A) The stopping of the motor using an external source.
B) Small movement of a motor.
C) Stalling of a motor.
D) Single-phasing of a three-phase motor.

73) What is the synchronous speed of a 575V three-phase 60Hz four-pole motor?

A) 3,600 rpm. B) 1,800 rpm.
C) 1,200 rpm. D) 900 rpm.

74) The disconnecting means for a motor shall be located within sight and _____ metres away.

A) 3. B) 5. C) 7. D) 9.

75) Some single-phase motors have a capacitor in the start circuit. The purpose of the capacitor is to

   A) Limit the starting current of the motor.
   B) Improve the power factor of the circuit.
   C) Limit the running current of the motor.
   D) Improve the starting torque of the motor.

76)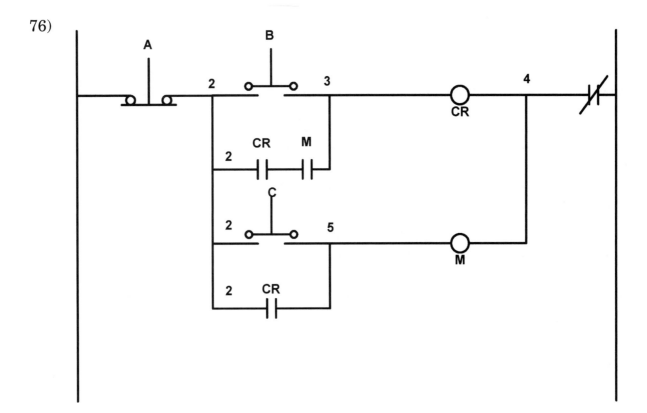

According to the diagram above, control relay CR will be energized and remain energized when

   A) Button A and C are pressed.
   B) Button C is pressed.
   C) Button B is pressed.
   D) Only if buttons B and C are pressed.

77) When additional stop buttons are placed in a control circuit, they must be wired

   A) In series with each other.
   B) In parallel with each other.
   C) In the power circuit.
   D) For safety reasons only 1 stop button is allowed in a control circuit.

78) What is the minimum size of time-delay fuse that can be used to protect a 10hp 500V DC motor?

   A) 25 A.   B) 30 A.   C) 34 A.   D) 40 A.

79) The following symbol represents

   A) A normally open, timed-open contact.
   B) A normally open, timed-closed contact.
   C) A normally open, held-open contact.
   D) A normally closed, held-open contact.

Block E: Extra Low Voltage Systems

80) What is the minimum size of conductor that can be laid in a raceway for a Class 1 extra-low voltage power circuit?

   A) #12 AWG.   B) #14 AWG.   C) #16 AWG.   D) #18 AWG.

81) When wiring a signal circuit for a supervised fire alarm system, you will be connecting

   A) Horns, pull stations, heat detectors, end-of-line resistor.
   B) Horns, flow switches, smoke detectors, end-of-line resistor.
   C) Horns, bells, strobe lights, end-of-line resistor.
   D) Horns, rate-of-rise detectors, end-of-line resistor.

82) What is the maximum length in metres of an emergency lighting system that is 12 V and is required to supply 8 lamps that are 6 W each with a maximum of 5% voltage drop when using a #10 AWG conductor?

   A) 36.5 m.   B) 29.4 m.   C) 22.9 m.   D) 18.7 m.

83) In order for a fire alarm system to have electrical supervision, devices must have what number of terminal connection points?

A) 2.   B) 4.   C) 6.   D) 8.

84) Emergency systems used to provide lighting and other emergency services in the case of failure of normal power are to be tested at least once a

A) Year.   B) Month.   C) Week.   D) Day.

85) When the trouble buzzer is ringing on a supervised fire alarm system, it means that

A) A pull station has been activated falsely.
B) The end-of-line resistor is of the wrong value.
C) A wire has come loose from its connection.
D) An automatic alarm-initiating device has been activated.

86) The maximum voltage that can be classified as "voltage extra low" is

A) 12 V.   B) 24 V.   C) 30 V.   D) 120 V.

87) When communication conductors are direct-buried in a trench with other direct-buried systems and the communication cables do not have a metal sheath, the minimum separation between the communication cables and the other systems must be

A) 300 mm.   B) 600 mm.   C) 900 mm.   D) 1 m.

88) A supervised fire alarm system that is labelled CLASS-A will have the end-of-line resistors

A) In the device box that connects the last device on the circuit.
B) In the control panel.
C) In the annunciator panel.
D) In a separate box after the last device on the circuit and labelled EOL.

89) A flow transmitter has an input range of 40 gallons per minute (GPM) to 130 GPM, using a standard output signal of 10 to 50 mA. If the output signal is 27.8 mA, what is the flow rate?

   A) 60 GPM.   B) 80 GPM.   C) 100 GPM.   D) 120 GPM.

Block F: Upgrading, Maintenance, and Repair

90) The transformer illustrated below is a 10:1 step-down transformer and is connected to a 120V supply. What value of volts will the voltmeter indicate when it is connected in this manner?

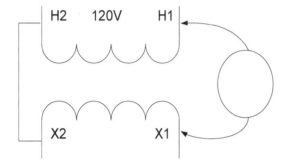

   A) 24 V.   B) 132 V.   C) 12 V.   D) 108 V.

91) When dealing with a PLC, the input/output (I/O) module is the part that

   A) Interfaces between the CPU and field-wired devices.
   B) Supplies AC power to the DC power supply.
   C) Rights the PLC programs.
   D) Provides surge protection from inductive loads.

92) When retrofitting an existing panel with a new bus and new moulded-case circuit breakers, your main concern is that

   A) The panel has room for future load growth.
   B) The panel has at least 25% spare breakers.
   C) The breakers have adequate fault current levels.
   D) The existing lines are long enough to reach the new bus bare connections.

93) When conducting an insulation resistance test using the 1,000V scale of a megohmmeter on a 575V motor, the test should be conducted for a minimum of

   A) 15 minutes.   B) 1 minute.   C) 25 seconds.   D) 10 seconds.

94) Upon inspection of a residential fuse panel, you find that a fuse has a blackened face. The most likely cause is that

   A) The fuse interrupted a short circuit.
   B) The fuse interrupted an overloaded circuit.
   C) The fuse opened due to arcing at the fuse holder.
   D) The fuse did not have the correct rating.

95) When using a clamp-on ammeter to check the current of a three-phase four-wire AC circuit, the meter must be

   A) Clamped on each current carrying conductor alternately.
   B) Clamped on the 3 current carrying conductors simultaneously.
   C) Clamped on all 4 conductors simultaneously.
   D) Clamped on the neutral conductor only.

96) The rating of an incandescent lamp is 130V and 2,000 hours. When the lamp is used on a 120V supply, the lamp will

   A) Seem dimly lit.   B) Seem brighter.
   C) Have a reduced lamp life.   D) Have an extended lamp life.

97) A silicon-controlled rectifier (SCR) is typically used in electronic circuits to

   A) Turn on the circuit at a predetermined time in the AC sine wave.
   B) Rectify the AC voltage to DC voltage.
   C) Regulate the current in the circuit.
   D) SCRs are not used in electronics.

98) During a service call, a residential customer says that when an appliance is plugged into a kitchen receptacle some lights in the house will dim while others will seem brighter. The most likely cause of the problem is

A) A ground fault on the kitchen circuit.
B) A bad neutral connection.
C) Out-of-date knob and tube wiring.
D) An over-current device in the neutral conductor.

99) If a hot water tank that is rated for 5,000 watts at 240 V AC single-phase was hooked up incorrectly to a 120V supply, the power of the circuit would be

A) 5,000 W.   B) 2,500 W.   C) 1,250 W.   D) 0 W.

100) You are performing a field-assembled connection between an aluminum lug and a copper bus bar using a 10mm bolt. To allow for expansion of the connection you must also include

A) 2 flat washers.
B) A conical spring washer.
C) 2 flat washers and 2 nuts.
D) A joint compound.

Answers to Practice Exams — 7

Answers to Practice Exam 1

1.	D	26.	C	51.	D	76.	D
2.	C	27.	D	52.	A	77.	B
3.	D	28.	B	53.	C	78.	B
4.	B	29.	B	54.	B	79.	D
5.	B	30.	A	55.	C	80.	B
6.	A	31.	C SEE HB p. 45	56.	C	81.	A
7.	B	32.	C	57.	B	82.	C
8.	C	33.	D	58.	D	83.	A
9.	D	34.	A	59.	D	84.	D
10.	A	35.	D	60.	C	85.	B
11.	C	36.	C	61.	B	86.	C
12.	B	37.	D	62.	A	87.	C
13.	B	38.	A	63.	C	88.	D
14.	D	39.	A	64.	B	89.	D
15.	D CEC 2-308	40.	C	65.	C	90.	A
16.	C	41.	D	66.	A	91.	A
17.	A	42.	C	67.	B	92.	D
18.	B	43.	B	68.	D	93.	C
19.	D	44.	D	69.	D	94.	B
20.	A CEC 8-102	45.	B	70.	B	95.	C
21.	C	46.	A	71.	C	96.	C
22.	D	47.	B	72.	D	97.	B
23.	B	48.	C	73.	C	98.	A
24.	A	49.	B	74.	A	99.	D
25.	A	50.	A	75.	A	100.	B

Answers to Practice Exam 2

1.	C	26.	C	51.	D	76.	C
2.	B	27.	A	52.	C	77.	A
3.	A	28.	D	53.	B	78.	A
4.	A	29.	A	54.	C	79.	B
5.	C	30.	C	55.	B	80.	D
6.	B	31.	A	56.	D	81.	C
7.	D	32.	C	57.	A	82.	C
8.	C	33.	B.	58.	C	83.	B
9.	B	34.	D	59.	D	84.	B
10.	B	35.	D	60.	A	85.	C
11.	B	36.	B	61.	D	86.	C
12.	A	37.	C	62.	C	87.	A
13.	B	38.	A	63.	D	88.	B
14.	B	39.	A	64.	B	89.	B
15.	D	40.	D	65.	A	90.	D
16.	A	41.	B	66.	D	91.	A
17.	D	42.	C	67.	B	92.	C
18.	B	43.	B	68.	C	93.	B
19.	C	44.	D	69.	B	94.	A
20.	B	45.	D	70.	C	95.	A
21.	B	46.	A	71.	D	96.	D
22.	D	47.	A	72.	B	97.	A
23.	B	48.	C	73.	B	98.	B
24.	A	49.	C	74.	D	99.	C
25.	C	50.	A	75.	D	100.	B

Explanations for Practice Exam 1 Answers 8

Block A: Occupational Skills

1) Answer D is correct.

On a job site there is a chain of command, and the first person in that chain, and the first person you should contact, is your foreman. The foreman will consult with the general contractor. Since the foreman is not one of the answers, the next person in line is the general contractor.

2) Answer C is correct.

The Occupational Health and Safety Act dictates that a worker's concerns are addressed by the owner or their representative and someone appointed by their peers as the on-site health and safety representative.

3) Answer D is correct.

The smaller the sling angle, the lower the working load limit. For example, if you have a 1,000 kg load and you connected 2 slings to a straight bar to lift the load, the slings would be at a 90° angle and would split the load evenly at 500 kg each. At a 60° angle, the load on each sling increases to almost 600 kg each, and at a 30° angle the load on each sling increases to 1,000 kg on each sling. The source for this answer is the Construction Safety Association of Ontario (CSAO). There will be a similar association in your province or territory.

4) Answer B is correct.

Hacksaw teeth point away from the handle because the cutting action happens on the downstroke. Most blades have an arrow on the blade and the word HANDLE, which will tell you which way the blade should be facing when installed. But the paint may be removed from a used blade, leaving the decision up to the user.

5) Answer B is correct.

The OHSA states the maximum distance is 1 unit out for every 3 units up, and the minimum distance is 1 unit out for every 4 units up. Answer B is the only one that falls between the two.

6) Answer A is correct.

A high-pot tester is short for a "high-potential tester." It is used to test the insulation of a high-voltage cable. The other 3 choices do not test voltage—not directly. A clamp-on ammeter is used to test the current of a conductor without breaking the circuit. A watt-hour meter is most commonly used to measure the power consumed in a building. A Wheatstone bridge is used to accurately measure resistance.

7) Answer B is correct.

Only the person who installed the lock is allowed to remove it.

8) Answer C is correct.

All MSDS sheets are to be placed in the on-site MSDS folder so anyone can view the potential hazards.

9) Answer D is correct.

The Trades Qualification Act states anyone doing work in a certified trade must apply for apprenticeship immediately. You have 3 months to file your contract with the director.

10) Answer A is correct.

The CEC is only interested in proper wiring methods. The Fire Code mainly determines the zoning of a building. And last, the ULC Standard determines the spacing requirements for devices.

11) Answer C is correct.

When time and material pricing is used the consumer is given a price per hour for labour and is to pay for any material. "Lump sum" is used when the pricing of a job is straightforward and many contractors may be bidding on a job. "Unit price" is used when the job is repetitive. "Paid when complete" is not a real term, only a distracter.

12) Answer B is correct.

There are 4 main categories of construction drawings: Architectural, Structural, Mechanical and Electrical. Open web steel joists are a structural component of a building, and so the depth of the joist would be found on a *structural* drawing.

Block B: Distribution and Services

13) Answer B is correct.

Delta or ungrounded systems must have ground-fault indicator lamps to indicate the presence of a ground fault because this will not cause an over-current device to trip. If a second phase were to be grounded, this could cause major damage to the electrical system. The ground-fault indicator lamps always fail to the OFF position. This indicates either a ground fault or a burned-out lamp.

14) Answer D is correct.

Rule 6-304 of the CEC states that the conductor size must be LARGER than #4. The only size of the 4 answers larger than #4 is #3. (Note: #5 is not a recognized electrical conductor size.)

15) Answer D is correct.

Rule 2-308 of the CEC states that the minimum head room is 2.2 m.

16) Answer C is correct.

See Rule 26-700 (8) (a), CEC.

17) Answer A is correct.

When current passes through a conductor, a magnetic field is created. With single-conductor cables in free air and with typical spacing (each cable is spaced 1 diameter apart from each other), this field is reduced to a tolerable level up to 200 A. See Rule 12-3022, CEC. 4-010 (4)

18) Answer B is correct.

The minimum size grounding conductor for any high-voltage installation is 2/0. See Rule 36-302, CEC.

19) Answer D is correct.

Trade knowledge: transfer switches can be automatic or manual, and are used to switch between normal and emergency power.

20) Answer A is correct.

A consumer's electrical system is broken into 3 parts: service conductors, feeder conductors, and branch circuit conductors. From the load side of a consumer's service to the point of utilization, a 5% voltage drop is allowed, but only a 3% drop for a feeder or branch circuit. See Rule 8-102, CEC.

21) Answer C is correct.

UPS is a backup or standby power source used when a loss of power, even for a short time, would cause major problems. UPSs use a battery, a charger, and an inverter to supply the required power needs.

22) Answer D is correct.

Examine the following calculation:

$50,000/(600 \times \sqrt{3}) = 48$ A

$48 \times 1.25 = 60$ A.

From Table 2 of the CEC, a #6 AWG conductor is required. See Rule 26-258, CEC.

23) Answer B is correct.

This only makes sense. On construction sites, heavy machinery may be used, and flexible cords that will be on the ground need to be able to handle the abuse. See Rule 76-002, CEC.

24) Answer A is correct.

Utilities deal with some of the highest voltages. When switching lines for maintenance or other reasons, the higher the load on the line, the higher the potential for a surge.

25) Answer A is correct.

See Table 2, CEC. Use the 85°C–90°C column, moving down until you find a conductor size rated for at least 60 A.

26) Answer C is correct.

The resulting kVA is 57.7% of the original 300 kVA, or 86.6% of the remaining 200 kVA.

27) Answer D is correct.

See Rule 2-400, or Table 65 of the CEC. Both of these explain the different categories of enclosures.

28) Answer B is correct.

The full-load amps of the capacitor is 24 A. See the calculation below:

$25,000 \text{ VA}/(600 \text{ V} \times \sqrt{3}) = 24 \text{ A}$

$24 \text{ A} \times 2.5 \text{ or } 250\% = 60 \text{ A}$.

Note: This is the maximum. The type of over-current device was not an issue; it would be the same for a fuse or circuit breaker. See Rule 26-210, CEC.

29) Answer B is correct.

The minimum size of grounding conductor is based on the maximum ampacity of the largest service conductor. See Table 17, CEC.

30) Answer A is correct.

For the bending radii of a 5kV portable power cable 25 mm to 50 mm in diameter, use a multiplier of 6 times the diameter of the cable. See Table 15, CEC.

31) Answer C is correct.

When a service mast or support member is above the roof line of a residence or building, it is no longer supported, and the pulling force from the utility incoming lines can put a major strain on the equipment. Rule 6-112 (Appendix B) of the CEC states that after 1.5 m a guy wire must be installed.

32) Answer C is correct.

When the insulation of a high-voltage cable is removed, the change in dielectric strength causes an electrical field to concentrate at the point where the insulation was removed. Stress cones are used to restore the dielectric strength and direct any voltage stress to ground.

33) Answer D is correct.

When the power factor of a building drops below 90%, a penalty on the utility bill is paid for wasted power.

34) Answer A is correct.

When disconnecting any switch that feeds a number of other loads, it is always a good trade practice to disconnect the downstream loads first. This will reduce any arcing that might occur due to the current through the switch at the time.

35) Answer D is correct.

Usually the metering is done on the load side of the service equipment. This is done for the safety of persons who may have to remove or replace a meter. Rule 6-402 of the CEC does make some exceptions to the rule. These fall mainly in the area of residential units, where the lower voltage and current rating make it reasonably safe and the problem of reading the meter is eliminated.

36) Answer C is correct.

Obviously minimum heights of cables must be set to ensure safety. Rule 6-112 of the CEC states the minimum requirements for various situations.

Block C: Branch Circuit Wiring

37) Answer D is correct.

Rule 12-012 and Table 53 must be consulted for the minimum coverage requirements for direct-buried cables.

38) Answer A is correct.

SO is a flexible cord. The conditions of use for SO can be found in Table 11 of the CEC, along with TEW. The other 2 wiring methods are cables, and their conditions of use can be found in Table 19. Rule 28-102 states that flexible cord may be used when connecting a portable motor, but it must have a rating of at least S.

39) Answer A is correct.

Rule 62-306 (1) (c) of the CEC states the cables must be 50 mm. The rest of the rule dictates the minimum depth of the concrete, not the cables. If the cables were buried deeper, the heat the cables produce would not be as effective. They must have a depth of 50 mm so that cuts may be made in the concrete without damaging the cables. The concrete must have a minimum depth so it will not crack and damage the cables.

40) Answer C is correct.

Trade knowledge: to control a light source from 2 locations, 2 three-way switches must be used. Any additional switches must be four-way switches. For example, if a light source was to be controlled from 100 locations, 2 three-way switches and 98 four-way switches would be used.

41) Answer D is correct.

Rule 12-616 of the CEC states that it must be 32 mm. The reason for this is the length of screws that the drywall installers use is 32 mm.

42) Answer C is correct.

Rule 12-502 of the CEC states that the maximum voltage is 300 V.

43) Answer B is correct.

Rule 12-1118, Appendix B, of the CEC gives the coefficient of linear expansion for a number of materials, including PVC. It also gives a sample calculation:

Expansion = length of the run in metres × the change in temperature × the coefficient of linear expansion.

Expansion = 40 m × 55°C × 0.052

Expansion = 114.4 mm.

44) Answer D is correct.

Table 23 and Rule 12-3034 of the CEC are used to solve the question. Table 23 states that ten #14 AWG conductors can occupy the box, but Rule 12-3034 states that a flush-mounted device on a single strap (switch or receptacle) takes up the room of 2 conductors. Also, the 2 wire connectors take up the space of 1 conductor, leaving space for 7 conductors.

45) Answer B is correct.

First, the reader must be clear on what a wireway is. A definition of a wireway can be found in Section 0 of the CEC. Here are the calculations:

75 mm × 75 mm = 5625 mm²

Rule 12-2104 (2) of the CEC states power conductors must be filled only 20%.

5625 mm² × 0.2 = 1125 mm². Table 10 of the CEC gives the area of the conductor.

1125 mm²/15.67 mm² = 71.8.

46) Answer A is correct.

A, B, and C require a ballast, but only A requires an igniter.

47) Answer B is correct.

Rule 62-214 (5) of the CEC states the distance that light sources or any other heat-producing equipment must be kept from an outlet box. This prescribed distance is set to protect against a possible fire hazard.

48) Answer C is correct.

This requires a calculation and the use of Table 8, 9, and 10 of the CEC.

The area required for a #3 R90 is 61.99 mm². The area required for 4 of these conductors is 247.96 mm².

The area required for a #4 R90 is 52.46 mm². The area required for 4 of these conductors is 209.84 mm². Total area is 457.80 mm².

Table 8 dictates that the maximum allowable conduit fill for 3 or more conductors that are not lead-sheathed is 40%. Table 9 shows a 41 mm conduit at 40% fill is 525.4 mm².

49) Answer B is correct.

Trade knowledge: steel shims will allow you to level the post on the base and tighten it firmly to the base, and the shims will not rot over time, which is a concern with an outdoor installation. Regarding answer A, adding some small amount of concrete to the top of the base will more than likely crack due to the fact the pole does move a small degree with outdoor conditions. Regarding answer C, there is likely no problem with the bolts, and installing new ones means removing the old ones. Answer D will do nothing to solve the problem.

50) Answer A is correct.

Heat pumps are usually used in residential applications to supplement the heating and cooling systems. They use a liquid line buried in the earth, extracting heat in the cooler weather and expelling heat from the building to the ground in the warmer weather.

51) Answer D is correct.

Rule 12-1010 (2) of the CEC states that if conduits of different sizes are placed on common supports, the supports shall be spaced according to the smallest conduit.

52) Answer A is correct.

An incandescent light source comes on instantly. Metal halide, high-pressure sodium, and mercury vapour, by contrast, are all high-intensity discharge (HID) light sources. The lamp can take 5 to 7 minutes to come to full brightness. In the case of a short power outage, the lamps must cool before a re-strike can occur. The cooling can take 10 to 12 minutes.

53) Answer C is correct.

Electric ranges have a space in the back to accommodate the large size of the moulded cord end. Keep the receptacle low and as close to the centre of the range as possible. See Rule 26-744 (5) (a), CEC.

54) Answer B is correct.

Lightning arresters are used to direct excess voltage from a lightning strike to ground. The voltage levels can be very high and place a great stress on the conductor. Keeping the conductor as straight and short as possible will help ensure the device operates correctly. See Rule 26-508, CEC.

55) Answer C is correct.

Rule 46-202 of the CEC states that the voltage level must maintain 91% of their rating. This will provide an acceptable lighting level to evacuate the building.

56) Answer C is correct.

Rule 12-3018 of the CEC gives the requirements for recessing boxes in combustible and non-combustible construction.

57) Answer B is correct.

Heat is the number one concern with conductors and must be looked at in 2 ways: the heat from current through the conductor, and the ambient temperature. If the ambient temperature is too high, the ampacity must be reduced. See Rule 4-004 and Table 5A, CEC.

58) Answer D is correct.

In Table D4 of the CEC, under the left-hand column marked CURRENT A, choose the current value that is closest to the actual circuit current. Then, move in a straight line to the right until you find a number equal to or greater than the 15 m that is required. Under the number 8 you see 16.8 m.

59) Answer D is correct.

Rule 18-108 of the CEC states that a seal shall be placed within 450 mm of any device that may produce sparks or high temperatures. In a Class I location, there are hazardous concentrations of gas or vapours that can migrate into a wiring system. Seals are used to ensure that a small explosion which might occur inside a device will not travel down that conduit or cable to an area that may cause a much larger explosion.

60) Answer C is correct.

Rule 12-3034 of the CEC states the calculations for determining the minimum dimensions for boxes that are used as pull or junction boxes that contain conductors that are #4 or larger.

Block D: Motor and Control Systems

61) Answer B is correct.

Adding external resistance to the shunt field circuit reduces the field flux. This causes a lower counter voltage. The reduced counter voltage causes a higher armature current and therefore a higher motor speed.

62) Answer A is correct.

A wye-delta starter uses 2 contactors to produce a "soft start." When the motor starts, the motor windings are in a wye configuration, giving the motor a lower line voltage. After a few seconds, the timing relay energizes the second contactor and de-energizes the first contactor connecting the motor windings in a delta configuration. This gives the motor full-line voltage.

If the delta voltage is three-phase 208, the starting voltage is $208/\sqrt{3} = 120$.

$120/208 = 0.577$.

63) Answer C is correct.

This should be the simplest of motor questions. Changing any 2 line leads will change the direction of the rotating magnetic field in the stator windings and therefore change the direction of the motor.

64) Answer B is correct.

Most circuit breakers are designed to provide both short circuit and overload protection. With motors having such a high starting or inrush current, a circuit breaker must be set too high to provide any overload protection. A separate overload device must be in place to provide this protection. Instantaneous trip circuit breakers do not have the thermal component, only the magnetic component that can react very quickly to a short or stalled motor.

65) Answer C is correct.

Until a number of years ago, the only way to regulate the speed of an AC motor was to use a wound rotor motor that uses a resistor bank. The motors are expensive and the resistor bank wastes much of the energy in heat. The speed of a motor depends on the number of poles and the frequency. With the advancement of electronics, the frequency of a motor is easy to change. Variable frequency drives have become very popular when speed control is required.

66) Answer A is correct.

Thermal overload devices heat up as the current through them, and therefore through the motor, increases. If you try to reset the device before it has had a chance to cool, it will not rest. Answers C and D are distracters.

67) Answer B is correct.

When a motor is controlled by a magnetic contactor with a holding contact, it is referred to as low-voltage protection or three-wire control. The remote station requires a power conductor and 2 additional conductors from the start button to the holding contact. If the question stated that the wiring method was a non-metallic raceway or a cable, then a bonding conductor would be required and the number of conductors would be 4.

68) Answer D is correct.

After consulting Table 44, you know that the RLC, or *rated load current,* is 22 A. To find the maximum size of over-current device, consult Rule 28-708, which states that the over-current device must not exceed 50% of the LRC, or *locked rotor current,* which is 6 times the RLC. If a device falling within this limit will not allow the motor to start, it may be increased to 65% of the LRC.

The calculation is as follows: 22 A × 6 × 0.65 = 85.8 A. This value is a maximum. When looking for an over-current device, Table 13 can be used. The standard over-current device that doesn't exceed 85.8 A has a maximum of 80 A. All tables and rules are from the CEC.

69) Answer D is correct.

Trade knowledge: Type J fuses are time-delay fuses designed for motor starting applications. HRC fuses are high rupture capacity fuses, and could be any of the other three types. Type P fuses are thermal or ambient temperature sensitive fuses. Type R fuses are a standard fuse, with no time-delay feature.

70) Answer B is correct.

The insulation class of the motor will determine the conductor insulation temperature rating. This can be found from the name plate of the motor and Table 37. Even though the conductor insulation temperature rating may be a minimum of 90°C or higher, when looking up the ampacity of the conductor you must look in the 75°C column of the ampacity tables. (See Rule 28-104, CEC.) Conductors that supply the motor connection box act in a small way as a heat sink, drawing heat away from the motor. Rule 28-106 states that the minimum ampacity of conductors supplying a motor shall be 125% of the motor's FLA. The calculation follows: 82 A × 1.25 = 102.5 A. Using the 75°C column of Table 2, the answer is a size #2.

71) Answer C is correct.

This type of motor has windings on the rotor unlike a squirrel cage motor. The secondary windings are connected to slip rings, and the slip rings to secondary conductors, and the secondary conductors to a resistor bank that is part of the controller. Rule 28-112 of the CEC states that the secondary conductors are to be sizes with the same multiplying factor as the primary conductors, which is 125%. The FLA of the motor is 52 A, according to Table 44 of the CEC. (52 × 1.25 = 65 A.) Using the 90°C column of Table 2, the answer is a #6 R90.

72) Answer D is correct.

As with all equipment that is CSA-approved, the manufacturer sets the guidelines for their equipment. To ignore them would void any warranty. The CEC is a guide, a last resort, you could say. If the manufacturer does not provide the information, Rule 28-202 of the CEC makes this clear.

73) Answer C is correct.

Table 25 of the CEC states that a motor connected to a single-phase supply requires 1 overload. A motor connected to a two-phase supply requires 2 overloads, and a motor connected to a three-phase supply requires 3 overloads.

74) Answer A is correct.

Generally with motors, the disconnecting means is to be located within sight and within 9 m. See Rule 28-604 (3), CEC. But because a number of other tradespeople besides electricians might service refrigeration equipment, the disconnecting means is to be within sight and within *3 m*, to ensure that these tradespeople have no difficulty locating the disconnecting means. See Rule 28-604 (5), CEC.

75) Answer A is correct.

Definitions of low-voltage release and low-voltage protection can be found in Section 0 of the CEC. With trade knowledge, this was an easy question. If you do not have experience in this area of the trade more study will be needed. Answer B and C are effectively the same; one cannot be more correct than the other. Answer D cannot be correct because a float switch is not controlled manually.

76) Answer D is correct.

The smallest size of over-current device required for power and lighting loads is 15 A. If 2 motors are small enough that their combined FLA does not require an overcorrect device greater than 15 A, they may be grouped together. See Rule 28-206, CEC.

77) Answer B is correct.

Motor conductor insulation temperature ratings, like almost all conductors, are based on an ambient temperature of 30°C. If the ambient temperature surrounding the motor is higher than 30°C, the insulation rating of the motor conductors must be increased to handle that exact temperature increase.

78) Answer B is correct.

The motor's FLA can be found in Table 44 of the CEC. The current is 62 A. Rule 28-06 states that this value is to be multiplied by 125%. This value is 77.5 A. Finally, in Table 2 you find the minimum size of conductor is a #4 R90 (XLPE).

79) Answer D is correct.

The formula for the speed of this type of motor is $S = 120 \times F/P$, where S is speed, 120 is a constant, F is the frequency, and P is the number of poles.

Block E: Extra Low Voltage Systems

80) Answer B is correct.

Fire alarm systems today are supervised. By having a resistor at the end of the circuit, the control panel will recognize a supervisory current. The end-of-line (EOL) resistor controls the level of this current. The calculation follows:

$3,900 \, \Omega / 24 \, V = 0.006 \, A$.

81) Answer A is correct.

See Rule 54-102 (1) (a), CEC. This rule gives the maximum open circuit voltage for a consumer's cable connection.

82) Answer C is correct.

Rule 54-202 of the CEC states that Class 2 conductors may be installed in the same raceways as conductive optical fibre cables. Class 2 circuits (Section 16) are low-energy circuits and pose a small threat to the optical fibre cables.

83) Answer A is correct.

All signalling devices for a fire alarm system have a diode in them, whether they are a bell, horn, or strobe light, and all fire alarm circuits must have an end-of-line resistor. Also, only signalling devices can be on a signal circuit. Answer B has no end-of-line resistor. Answer C has a diode reverse polarity. And answer D has a pull station which is not a signalling device in the circuit.

84) Answer D is correct.

If the wires for the window contacts are connected in parallel, then all of the contacts must be open for the alarm to sound. If the connections are made in series, any window contact will break the circuit and the alarm will sound. The other two answers are only distracters.

85) Answer B is correct.

Fire alarm circuits carry a very small current; therefore, their size can be smaller than power and lighting circuit conductors. But pulling conductors into a raceway does put a strain on them, so a minimum size is not based on current draw but on the integrity of the conductor. Rule 32-100 of the CEC gives the limitations for all installations of conductors in a fire alarm system.

86) Answer C is correct.

Type CFC conductors are a type of FCC, or flat conductor cable. Although not widely used, they do come in power, data, and communication conductors. CFC conductors are part of a system sometimes called "under carpet wiring." This system is used under carpet tiles and eliminates the need for jiffy poles or drops of any kind. Rule 12-808 of the CEC.

87) Answer C is correct.

Answers A, B, and D are examples of hardware, or the physical (software is not really physical, only code and programming contained within the memory of a physical medium, like a CD-ROM or hard drive—both examples of hardware) component that uses power from the computer power supply or the wall socket to operate.

88) Answer D is correct.

Two-stage fire alarm systems are used in places where it would be an extreme inconvenience to evacuate a building for a false alarm, such as a hospital. These systems wait for 5 minutes before going into evacuation mode. The intent of the 5-minute delay is to allow authorized, trained staff to first investigate the possibility of a false alarm before a very difficult or inconvenient evacuation is undertaken after this time has elapsed.

89) Answer D is correct.

All of these areas are places where patients could find themselves in need of a nurse's assistance.

Block F: Upgrading, Maintenance, and Repair

90) Answer A is correct.

A three-phase three-wire system is also called an ungrounded, or delta, system. Because the system is ungrounded, a ground fault will not cause an over-current device to open and the system will function normally. This is why such a system requires ground-fault indicator lamps. If 1 phase is grounded, that phase will be at the same potential as the non-current carrying metal it has been grounded to. Therefor, the voltage of the other 2 phases to ground will be the same as the voltage of the other 2 phases to the phase that has been grounded.

91) Answer A is correct.

Galvanic corrosion is a common process like rust that happens when 2 dissimilar metals are in contact with each other in air. Many times copper and aluminum must be connected in the field to prevent this process from happening. An aluminum joint compound is used over the connection that keeps the air out and eliminates the corrosion.

92) Answer D is correct.

One of the formulas for finding resistance in a parallel circuit is R1 × R2/R1 + R2. This is 18/9 in this circuit, and the answer would be 2 Ω. Remember the resistance in a parallel circuit is always less than any single resistance in the circuit. In this question, the only answer that had a value that was lower than the lowest resistance value in the circuit was D. Knowing the rules for series, parallel, and combination circuits can sometimes eliminate the need for a calculation, saving time for other more difficult questions.

93) Answer C is correct.

In instrumentation, 4 to 20 mA is a common signal that is used for process control. To answer the question, we have 16 divisions between 4 and 20. By dividing 1,200 V by 16 divisions we have each division equaling 75 V. 75 V multiplied by 3 divisions equals 225 V. Remember, you cannot multiply the 75 V by 7 mA because the signal level does not start at 1 mA; it starts at 4 mA. Therefor, 7 represents 3 divisions.

94) Answer B is correct.

All control circuits have inputs such as push buttons and outputs such as coils and lights. Output mismatch could not be correct because the program has been working for some time. The CPU not working also cannot be true because the program is running. The counter is a distracter. Of the 4 choices, it must be that an input is not functioning and therefor not allowing a rung to come true.

95) Answer C is correct.

Thermo photography is done with an infrared camera to catch any hot spots so that they can be repaired or replaced early, before a larger and much more costly repair is needed later. It is a preventative maintenance procedure. For this procedure to be most effective, all or as many loads as possible should be on.

96) Answer C is correct.

Volt meters measure a difference in potential. If a fuse is good, then there should be no difference in potential from one end of the fuse to the other. The meters testing L1 and L2 show zero volts. This does not mean there is no voltage that could be applied, it means there is no difference in potential between these 2 points. The meter testing L3 does show a difference from one end of the fuse to the other. This fuse must be open.

97) Answer B is correct.

The first thing to notice is the last colour; this gives the tolerance. If the colour is *silver*, the tolerance is 10%, but since it is *gold*, which is 5%, answer C and D must be incorrect. Next, the value of the remaining colours: *red* gives a first digit of 2, *blue* gives a second digit of 6. Next, *orange* gives the number of zeros after the 2 digits. The value is 26,000 Ω.

98) Answer A is correct.

Different types of batteries should be handled according to the manufacturer's guidelines. Lead acid batteries should be well known to the electrician; they are used for emergency power on many occasions. A lead acid battery should always be kept fully charged; a discharged battery can cause sulfation, which is an insulating layer that forms on the plates. This can be irreversible, at which point the battery is not useful any longer.

99) Answer D is correct.

Rule 12-904 of the CEC states that all conductors of a circuit must be contained in the same metal raceway. The reason for this is that as current passes through a conductor, a magnetic field is created and will affect any ferrous metal within this field. When all conductors of a circuit are in close proximity to one another, the field of each conductor helps to cancel out the other field. Example: if you measure current of 1 line of a three-phase motor while running, you will receive an accurate reading of the current in that line. But if you measure all 3 lines at once, the reading will be zero. This is of course because the different fields are cancelling each other out.

100) Answer B is correct.

Wound rotor motors are still used in many application; cranes are probably the most commonly used application. For this application, speed control and a high starting torque make these motors ideal. But when upgrading, a VFD will take the place of the resistor bank used for the wound rotor motor. And the motor itself must be changed as well. VFDs can provide speed control from 1% to 100% with a very smooth transition.

Explanations for Practice Exam 2 Answers 9

Block A: Occupational Skills

1) Answer C is correct.

Rule 2-004 of the CEC states that a permit must be obtained before work begins.

2) Answer B is correct.

The Trades Qualification Act dictates that it is the responsibility of the apprentice to keep track of their own hours.

3) Answer A is correct.

Like most codes or acts, this is a minimum standard. Companies will regularly impose stricter guidelines, especially when it comes to safety.

4) Answer A is correct.

The answer to this question can be found in a number of places—the National Building Code or CEC to name 2. Anyone in charge of ordering material should know that FT1 is for combustible construction and FT4 is for non-combustible construction. In certain parts of the country, FT6, or plenum-rated cable, must be used when installing cables in a drop ceiling or other place that is used as a return air plenum.

5) Answer C is correct.

The Occupational Health and Safety Act states that the minimum age of a person working on a job site is 16. Furthermore, the Trades Qualifications Act also states that the minimum age for an apprentice is 16.

6) Answer B is correct.

Trade knowledge: any apprentice or journeyman should be aware that every electrical room should have in a frame and under glass, or plexi-glass, a single-line diagram showing the major components of the building's electrical system, and the order in which they appear. A plan view looks directly down at the object, like a floor plan or site plan. An elevation looks straight on at an object, such as the outside wall of a building; and a shop drawing is something that is usually added to a set of drawings by a manufacturer of a specific construction material.

7) Answer D is correct.

WHMIS training is a must for all workers on a construction site.

8) Answer C is correct.

Scaling of a drawing is a necessary tool that electrical workers need to be able to use. With metric measurements, the length of the line, assuming millimeters, is multiplied by the scale and divided by 1,000. Thus: $75 \times 20/1,000 = 1.5$ m.

9) Answer B is correct.

These are common hoisting and rigging hand signals that can be found in many construction safety manuals.

10) Answer B is correct.

When performing work on service entrance equipment, your meter must be CAT 4. Downstream of the service entrance equipment, CAT 3 meters are sufficient.

11) Answer B is correct.

Trade knowledge: If the other trades are doing the finishing, then the electrical trade should also be at the same stage. Fixtures, receptacles, switches, and cover plates are all part of the finishing.

12) Answer A is correct.

All provinces have a construction safety association and usually have a safety manual for each individual trade. Be sure that you are in possession of the one for your area. The recommendation is that debris be removed on a regular basis, and no longer than once a day.

Block B: Distribution and Service

13) Answer B is correct.

Rule 62-116 (2) (a) and (b) of the CEC state that 100% of the first 10 kW and 75% of the remainder of the heating load be added to the service calculation.

14) Answer B is correct.

Table 36A of the CEC is used for aluminum neutral supported cable and Table 36B is used for copper. NS75 is the insulation temperature rating and quadraplex means there are 4 conductors: 3 phase conductors and a bare neutral that supports the other 3.

15) Answer D is correct.

Table 2 of the CEC is consulted first to determine the size of conductor required for the 100A service. This is a #3 R90 conductor. Three conductors are required for a 120V/240V circuit. Then, Table 6 is used to determine the minimum conduit size.

16) Answer A is correct.

Using Table D12b, Detail 3, which has 3 conductors per phase, the service ampacity must be divided by 3, giving us 333 A. Using 333 A, the minimum conductor size is 350 kcmill.

17) Answer D is correct.

The basic load is based on area given—5 kW for the first 90 m^2 and 1 kW for each additional 90 m^2 or part thereof. Therefor, the basic load is 6 kW. See Rule 8-200 of the CEC for single dwellings.

The electric range is 6 kW for any range up to 12 kW. The pool heater and air conditioner are specially noted at 100% of the load. The water heater is not a tankless model and is therefor not noted, because the water heater is in excess of 1,500 W and because an electric range has been used, the load is taken at 25%. The calculation follows: 6 kW + 6 kW + 1 kW + 4 kW + 5 kW = 22 kW.

18) Answer C is correct.

Rule 36-302 of the CEC states that an outdoor station (pad-mount transformer would be considered a station) requires a minimum of 4 ground rods. Note: Consult the definitions at the beginning of Section 36.

19) Answer C is correct.

Rule 4-008 (Appendix B) of the CEC states that in order to eliminate sheath currents you should ground the supply side and isolate by using a non-metallic material (usually a fibre board) on the load side.

20) Answer B is correct.

When dealing with high-voltage installations, we must be concerned about the possibility of touch and step voltages. Consult the definitions at the beginning of Section 36.

21) Answer A is correct.

An inverter may have all of the other 3 devices included within it, but the inverter is a single unit.

22) Answer B is correct.

Rule 6-104 of the CEC states that the maximum number of services on a building is 4. The most common application of this is row housing, but there are others.

23) Answer B is correct.

Section 26 deals with the construction requirements for a fence if it is needed to protect the average person from the dangers associated with electrical equipment that has open live connections. Rule 26-308 (4) states that the maximum spacing is 3 m.

24) Answer A is correct.

To gauge the approximate value of the available fault current at the secondary terminals of a transformer, divide the secondary current by the % impedance. Thus:

150,000 VA/(208 × √3) = 417 A.
417A/0.05% = 8,340 A.

25) Answer C is correct.

When using an ungrounded (or delta) supply, a ground fault will go unnoticed until a second phase shorts to ground, potentially causing considerable damage to the electrical system. For this reason, ground-fault indicator lamps are the most common method used to indicate the presence of a ground fault. See Rule 10-106 (2), CEC.

26) Answer C is correct.

Rule 8-110 (c) of the CEC states that the areas of the first or second floor of a dwelling must be considered at 100%. The basement has fewer loads than the other floors and can therefore be considered at 75%.

27) Answer A is correct.

Many people think it is the electrical inspector or the code book that determines the location of service equipment, but it is the supply authority. See Rule 6-116 (a), CEC.

28) Answer D is correct.

You must consult Rule 8-210 and Table 14 of the CEC. The basic load for a building listed in Table 14 is found by using the area in meters squared based on outside dimensions, and then using the multiplier for watts per meters squared. Next, use the demand factor in Table 14 for service or feeder conductors, depending on what you are using the calculation for. Here is the calculation:

Area: $35 \text{ m}^2 \times 20 \text{ m}^2 = 700 \text{ m}^2$
$700 \text{ m}^2 \times 5 \text{ W} = 3,500 \text{ W}$
$3,500 \text{ W} \times 0.7 \text{ demand} = 2,450 \text{ W}$.

29) Answer A is correct.

Service equipment that is above a certain voltage or current rating is required to provide ground-fault protection. See Rule 14-102 (1) (a), CEC.

30) Answer C is correct.

Series-rated combinations provide protection for downstream circuit breakers that have a fault level that is less than the available fault current of the circuit. These breakers will operate normally if an overload condition exists, but the upstream circuit breaker will trip first if a fault condition exists. This is not selective coordination, and can make troubleshooting more difficult. See Rule 14-014, CEC.

31) Answer A is correct.

Rule 8-108 (1) (b) of the CEC lays out the minimum number of branch circuits required for all standard panel boards used in dwelling units. If the question pertained to an apartment, you must read Sub-rule 3.

32) Answer C is correct.

There are a number of different methods that can be used to ground a service. Ground plates have become popular because only one is needed, and they are much easier to install than other types of grounding electrodes. See Rule 10-700 (2) (b), CEC.

33) Answer B is correct.

To find the load on the service equipment, you must consult Rule 8-202 of the CEC.

	Unit A	Unit B
Area	47 m² → 5,000 W	43 m² → 3,500 W
Electric range	11 kW → 6,000 W	10 kW → 6,000 W
Electric heat	8 kW → 8,000 W	7 kW → 7,000 W
Total wattage	19,000 W	16,500 W

Once you have found the total wattage for each unit, to find the service calculation you must remove the electric heat value from each unit. (It will be added back in later using Rule 62-116 of the CEC.) Next, using 100% of the load of the unit having the heaviest load and 65% of the next unit, the electric heat will be added in for both at 100% for the first 10kW and 75% for the remainder of the electric heat load.

Unit A at 100% =	11,000 W
Unit B at 65% (9,500 W × 0.65) =	6,175 W
Electric heat, first 10 kW at 100% =	10,000 W
Remainder of electric heat load at 75% (5,000 W × 0.75) =	3,750 W
Total wattage is	30,925 W

The calculated load on the service equipment in amps is the total wattage divided by the system voltage. 30,925 W/240 V = 128.8 A.

34) Answer D is correct.

Table 17 of the CEC gives the minimum size of grounding conductors for services. The conductor size is based on the ampacity of the ungrounded conductor. For any service over 800 A, the conductor size is 2/0 AWG.

35) Answer D is correct.

Tap changers are found on many transformers. The large transformers used by the utilities will have tap changers that may be changed under load, but most transformers cannot be changed under load. The main purpose of the tap changer is to achieve the proper secondary voltage when the primary voltage is slightly different than the supply voltage.

36) Answer B is correct.

Lightning arresters are used to direct excess voltage to ground before it can cause damage to the building electrical system. The voltage and current values of lightning can be extreme; therefor, the shorter the distance, the less chance there is of a problem. See Rule 26-508 (2), CEC.

Block C: Branch Circuit Wiring

37) Answer C is correct.

Table 2 will be used to determine the ampacity for up to 3 conductors in a cable or raceway. The ampacity for a #6 AWG conductor is 65 A. Table 5C is then used to derate the conductors because there are more than 3 current carrying conductors in the raceway. The derating factor for 4 conductors is 80%. Next, Table 5A must be used to further derate the conductors due to the high ambient temperature. The deration factor in Table 5A for 90°C conductors is also 80%. 65 A × 0.8 × 0.8 = 41.6 A. Now, with this calculation it does not matter which table is used or in what order as long as both are used.

38) Answer A is correct.

Rule 12-3034 (4) states that in order to find the number of conductors allowed in a box, you must find the wire size from Table 22 and use the area required for each conductor divided by the space of a given box. Thus: 775 ml/36.9 ml = 21 conductors.

39) Answer A is correct.

Using Table 6 of the CEC, the answer is 27 mm. Note: Be sure you are looking at the correct insulation type when using Table 6—the conduit sizes can change quite a bit. Using the incorrect page will result in a wrong answer to an easy question.

40) Answer D is correct.

Bonding conductors are allowed to be un-insulated under these conditions. If they are any longer, or have more bends in the run, it could result in the bonding conductor damaging the insulation of the insulated conductors.

41) Answer B is correct.

This question can be tricky. Most people think of a receptacle when referring to GFCI protection. Rule 68-064 states that receptacles cannot be closer than 1.5 m to the water's edge and that receptacles 1.5 m to 3 m from the edge must be GFCI protected. But the GFCI protection cannot be closer than 3 m. The reason for this is that in order to reset the GFCI protection, a person must get out of the pool water. See Rule 68-068 (6) (b), CEC.

42) Answer C is correct.

Ohm's Law: I = E/R. If the voltage is doubled, the current will double. If the resistance is reduced to half, the current will double. If both conditions occur, the current will increase by 4 times.

43) Answer B is correct.

First, you must be aware of the neutral and the ungrounded conductor location on the receptacle. The longer line on the receptacle is the neutral, and the shorter one is the live or ungrounded conductor. The connection between the two screws for the ungrounded conductors must be removed. One conductor for constant power would be connected to the bottom screw, and one conductor controlled by the switch would be connected to the top screw.

44) Answer D is correct.

In a three-phase delta system, the line current is equal to the phase current multiplied by the square root of 3. IL = I ph (phase) × $\sqrt{3}$.

45) Answer D is correct.

PVC conduit is not metal and if it is subjected to excessive temperatures, it will begin to expand and warp. See Rule 12-1104 (1), CEC.

46) Answer A is correct.

Incandescent light sources are the most inefficient, with up to 80% of the energy used lost as heat energy. Fluorescents are the next group, much more efficient than incandescent but not nearly as efficient as some of the high-intensity discharge light sources. Although low-pressure sodium is a low pressure light source, it is usually lumped in with the HID light sources, and is by far the most efficient light source. Low-pressure sodium is not widely used though because of its poor colour rendering. There are now colour corrected low-pressure sodium lamps that are starting to make inroads into more common use.

47) Answer A is correct.

This is done to protect people. If the heating cables are not totally embedded, there is a risk that they could be damaged, which could result in an injury to someone. See Rule 62-114 (5), CEC.

48) Answer C is correct.

The answer to this question can be found in 2 ways. If you are familiar with the configuration of the receptacle, Diagram 1 of the CEC could be used (diagrams follow the Tables section in the CEC book); if not, then you can use Rule 26-744 (2). Remember, if the designation for the device has an L in front of it, this is a *locking* type device. If the device has a P at the end of the designation, the device is a *plug* and not a receptacle.

49) Answer C is correct.

Flexible cords are not the same as wires and cables; they are not rated for the same ampacity. Table 12 of the CEC must be consulted for the ampacity of flexible cords. The ampacity of a #8 AWG flexible cord with 3 current carrying conductors is 35 A. Next, look under the title of the table and see Rule 4-014. This rule gives the derating factors for cord found in Table 12. The factor is 80%, so the calculation is: $35 \times 0.8 = 28$ A.

50) Answer A is correct.

Rule 26-750 (1) of the CEC states the secondary protection is set to 96°C. The secondary protection is the high limit. If you choose 90°C, you were reading Sub-rule 2, which describes the thermostatic control.

51) Answer D is correct.

Rule 20-004 (3) of the CEC states the area classifications for a gasoline dispenser and the surrounding area.

52) Answer C is correct.

The function of a ballast is to step up the voltage to strike the arc necessary to excite the gas in the lamp, after which the ballast must limit the current or the light source would quickly burn out. A neon light source does not use a ballast; it only needs a transformer. Beware when dealing with neon light sources because the voltage levels can be extremely high. For such sources, 15,000 V is not uncommon.

53) Answer B is correct.

Rule 12-214 (1) (a), CEC. States the minimum thickness of running board is 19mm. One of the more common materials used for a running board is lumber. A 1" x 4" board is really 3/4" × 3¾," with 3/4" converting to 19 mm.

54) Answer C is correct.

EMT is not as strong as ridged metal conduit and therefore must have more restrictive supporting methods. See Rule 12-1404, CEC.

55) Answer B is correct.

Trade knowledge: a 4" × 1½" octagon box is a standard trade size box commonly used to hold light fixtures and 8/32 screws are used in these boxes. Here, 8 describes the diameter of the screw and the 32 is the number of threads per inch.

56) Answer D is correct.

Rule 12-3036 (2) (b) of the CEC states that when using a pull box for a straight pull, the minimum distance between the 2 raceways must be at least 8 times the largest raceway if they are of different sizes. Thus, 53 mm × 8 mm = 424 mm.

57) Answer A is correct.

Panel boards in dwelling units should be easily accessible, basically about eye level. See Rule 26-402 (2), CEC.

58) Answer C is correct.

The voltage ratings for some wires and cables can be found in a rule, in this case Rule 12-502. Most of the time, you would do better to look at Table D1 of the CEC.

59) Answer C is correct.

ELC is extra low voltage cable (also stands for "extra low voltage control cable"). Table 19 and Table D1 can be used to find information on wires and cables. The answer to this question was found in Table 19.

60) Answer A is correct.

Trade knowledge: an isolated ground system requires an insulated bonding conductor that is attached to the receptacle and a separate bonding conductor connected to the non-current carrying metal.

Block D: Motor and Control Systems

61) Answer D is correct.

Trade knowledge: the normally open (N/O) contact maintains the power to the coil when the start button is released. But it is also commonly referred to as a "holding" or "sealing" contact.

62) Answer C is correct.

A series motor must always be directly connected to the load due to the fact that this type of motor will continue to increase speed to a dangerous level within a few seconds of being disconnected from the load. Series motors do not have a predetermined top speed and will increase in speed until the motor self-destructs.

63) Answer D is correct.

Rule 28-306 of the CEC gives 2 multipliers for determining the maximum value of motor overload protection. Use a multiplier of 1.15 if the marked service factor is less than 0.15 or is not marked, or 1.25 if the marked service factor is 1.15 or greater.

64) Answer B is correct.

A rule of thumb for 575V three-phase motors is 1 A per hp. A review of Table 44 of the CEC will show this principle. This principle holds true for other 575V or 600V three-phase electrical equipment. For example, a 20 kVA 600V three-phase transformer will have a current equal to approximately 20 A.

65) Answer A is correct.

Low-voltage protection is also called three-wire control. With this type of control circuit, a maintaining contact is used. If the voltage drops too low to keep the coil energized, the coil cannot be re-energized by the return of normal operating voltage unless the start button is pressed. This is the preferred control circuit when an unexpected starting of a device could cause harm to the operator. A definition of low-voltage protection and low-voltage release is in Section 0 of the CEC.

66) Answer D is correct.

Single-phase motors require a start winding to create the attraction and repulsion needed to create the turning force. Once the motor is turning, the start winding is no longer required and must be disconnected from the supply or else it will burn out. Without the start winding, the motor may not start.

67) Answer B is correct.

When used for motor starting applications, auto transformers are called a reduced-voltage starter. There are other methods of reduced-voltage starting called a wye-delta starter and a primary resistance starter. These starting methods reduce the voltage on start-up; therefor, the current is reduced. These are all forms of a "soft starting."

68) Answer C is correct.

There are a number of ways to change the direction of a direct current motor. The most accepted method is to change the armature connection.

69) Answer B is correct.

Trade knowledge: start windings are used on single-phase motors. The start winding uses a small gauge of wire compared to the run winding and must be disconnected as soon as it is no longer needed. This is at approximately 75% of motor-rated speed.

70) Answer C is correct.

The motor is no longer receiving the smooth-rectified direct current, but is now receiving a half-wave pulse. Depending on the loading of the motor, it will most likely slow. If the motor is fully loaded, it might not be able to overcome the load and will stop.

71) Answer D is correct.

Rule 28-200 of the CEC states that the value used is determined from Table 29 and that these values are the maximum. Once the calculated value is obtained, use Table 13 to determine the correct size of circuit breaker by rounding down to the next standard size.

72) Answer B is correct.

A jog button is like any start button, with the exception that the button must be held down to keep the circuit energized. There is no maintaining contact in this circuit.

73) Answer B is correct.

The formula for the speed of the magnetic field of a motor is $S = 120 \times F/P$. Where 120 is a constant, F is the frequency and P is the number of poles.

74) Answer D is correct.

Rule 28-604 (3) of the CEC gives the maximum distance from the motor that a disconnecting means must be.

75) Answer D is correct.

Single-phase motors do not have the advantage of a three-phase rotating magnetic field and require help with starting. Some single-phase motors, depending on what they are used for, may require extra help starting, and use a start winding and a capacitor.

76) Answer C is correct.

When button B is pressed, coil CR is energized and both CR contacts are closed. When the CR contact that is parallel to button C is closed, coil M is energized. At this point, both contacts CR and M are closed and coil CR will remain energized.

77) Answer A is correct.

Control circuits can have many stop buttons, and all must be wired in series with each other.

78) Answer A is correct.

The motor's FLA can be found in Table D2 of the CEC. The current is 18 A. Next, Rule 28-200 states that this value is to be multiplied by the factor found in Table 29. The maximum multiplier for direct current motors is 150%, which would give a value for this example of 27 A. According to Table 13, the standard size of time-delay fuse that doesn't exceed 27 A is 25 A.

79) Answer B is correct.

The symbol is a timing contact. Timing contacts are never held open or closed. The contact is normally open, due to the fact that the line is hanging down below the contact point. The arrow head at the end of the symbol represents when the timing will begin in relation to the coil. The arrow pointing *up* means that the timing starts when the coil is energized (or is picking up). The arrow pointing *down* means that the coil is being de-energized (or is dropping out).

Block E: Extra Low Voltage Systems

80) Answer D is correct.

When conductors are laid in raceways as opposed to being pulled into a raceway, the size of the wire can be decreased. See Rule 16-110 (2) (b), CEC.

81) Answer C is correct.

All circuits of a supervised fire alarm system must have end-of-line resistors at the end of the circuits. Pull stations, heat detectors, rate-of-rise detectors, smoke detectors, and flow switches are all alarm-initiating devices and cannot be on the same circuit with signal circuits. The only devices that can be on a signal circuit are devices that signal when an alarm condition occurs.

82) Answer C is correct.

Using Table D4 of the CEC, you must find the current of the circuit. Next, find the current value from the table that is closest to the actual current. Then, find the factor and the wire size and place the values into the formula given in the notes of the table. Thus: 8 lamps × 6 W = 48 W. 48 W/12 V = 4 A. The value from the table closest to this value is 4.25 A. The factor from the table, using 4.25 A and #10 AWG conductor size, is 10.8.

Therefor:

L = (12 V/6 V) × (5% / 5%) × (4.25 A/4 A) × 10.8
L = 22.95 m

83) Answer B is correct.

In order for a fire alarm system to be supervised, the conductors entering and leaving a device must be terminated separately. Otherwise, a conductor could come loose from a connection point and the control panel would have no signal that the resistance of the circuit had changed. Rule 32-106, CEC.

84) Answer B is correct.

To ensure that emergency systems will operate as expected, testing must be done at least once a month. Rule 46-102 (1), CEC.

85) Answer C is correct.

If A or D has occurred, the system will be in alarm mode and all signaling devices will be activated. The trouble buzzer of the fire alarm system will indicate if a wire has come off a device or if a ground-fault condition exists.

86) Answer C is correct.

Section 0 of the CEC has the definitions for voltage, extra low, low voltage, and high voltage.

87) Answer A is correct.

Rule 60-600 (b), CEC.

88) Answer B is correct.

There are 2 types of wiring for non-addressable supervised fire alarm systems Class A and Class B. Class A wiring has the EOL resistors in the control panel. This is a preferred wiring method for troubleshooting purposes. Class B systems have the EOL resistors in a separate box after the last device on the circuit.

89) Answer B is correct.

The range of the flow meter is 40 to 130 GPM and the difference is $130 - 40 = 90$ GPM. The transmitter has a range of 10 to 50 mA and the difference is $50 - 10 = 40$ mA. This gives us 2.25 GPM per mA ($130/40 = 2.25$).

Remember, the signal starts at 10 mA. If the signal is 27.8 mA, this is an increase of 17.8 mA from the base reading. $17.8 \text{ mA} \times 2.25 \text{ GPM} = 40$ GPM. This value plus the base reading of 40 GPM equals 80 GPM.

Block F: Upgrading, Maintenance, and Repair

90) Answer D is correct.

This type of connection is only used to determine whether the transformer has an additive or a subtractive polarity. This transformer has a subtractive polarity. With a ratio of 10:1, the secondary voltage is 12 V. Therefor, $120 \text{ V} - 12 = 108$ V.

91) Answer A is correct.

The I/O module is what the devices are wired to (outside world). Without the I/O, the PLC is just a computer without a purpose.

92) Answer C is correct.

Panels are often retrofitted with a new bus bar and new breakers because of the cost of the old, outdated breakers. Care must be taken to ensure that the new equipment has adequate fault current levels.

93) Answer B is correct.

When conducting an insulation resistance test, it is a common misconception that the test need only be conducted for a few seconds. The test should be conducted for a minimum of 1 minute because the wire of coil needs time to charge up like a capacitor. Once this happens, if the insulation is deteriorating the needle will deflect. If the insulation is good, the needle—or number, if using a digital model—will continue to rise.

94) Answer A is correct.

Plug fuses are the type used in older residential fuse panels. You can tell if the problem was a short or overload by inspecting the fuse. A short causes a high level of energy and will turn the face of the fuse black. If the problem was an overload, you may be able to see that the fuse link has opened and that the face is not black.

95) Answer A is correct.

A clamp-on ammeter only reads the strength of the magnetic field around the conductor and then turns that into a value of current. If conductors of different phases are checked at the same time, the opposite fields help to cancel each other out and the reading, if any, will be of no use.

96) Answer D is correct.

Incandescent lamps have a fairly short life, but by reducing the voltage to these lamps, the life can be extended.

97) Answer A is correct.

A silicon-controlled rectifier is a type of three-terminal thyristor that conducts current when triggered by a voltage at the gate terminal, and it remains on until the anode falls below a specified value.

98) Answer B is correct.

Neutral conductors have 2 purposes: they carry the unbalanced current, and they create a stable voltage. If the neutral connection is broken somewhere, the voltage will become unstable. This may not be seen easily with appliances. They will either work or not work, but with incandescent lamps some receive less voltage and will dim, while others on the other side of the phase will see an increase in voltage and will become brighter.

99) Answer C is correct.

A good understanding of Ohm's law will make this an easy question. One of the formulas for power is $P = E^2/R$. With this equation, it is easy to see that if you reduce the voltage by one half, you will reduce the power to one quarter. The other way to resolve the problem is to realize the one element of the circuit that will remain constant is the resistance. Find the resistance of the circuit under normal operating conditions and then apply this value to the formula $R = E^2/P$. Thus: $R = 11.52 \, \Omega$

$P = 120^2/11.52 \, \Omega$

$P = 120$ W.

100) Answer B is correct.

There are different types of field connections. Rule 12-118 (5) (a) of the CEC gives the requirements for this type of connection.

Recommended Study Texts  10

The following textbooks can be used to aid in the preparation for the C of Q exam. Some are used in colleges that prepare apprentices, some are used on the job, and some are of a more specialized nature. Some may be found in college bookstores while others can be found through online retail services. A brief description follows each book.

Canadian Electrical Code Book 2006, Part 1, 20th Edition
CSA Canadian Standards Association
ISBN 1-55436-023-4
Covers electrical installation guidelines based on fire and shock hazards and proper maintenance.

Applications of Electrical Construction, 3rd Edition
Robert K. Clidero, Kenneth H. Sharp
ISBN 0-7725-1719-3
Covers wiring methods, materials, and installation norms.

Electric Motor Controls for Integrated Systems, 3rd Edition
Gary Rockis, Glen Mazur
ISBN 0-8269-1-207-9
Covers motors, control methods, and circuits.

Delmar's Standard Textbook of Electricity, 3rd Edition
Stephen L. Herman
ISBN 1-4018-25656
Covers all relevant electrical theory at the required level.

Fire Alarm Systems—A Reference Manual
Canadian Fire Alarm Association
ISBN 0-9692433-0-8
An overview of fire alarm systems, design, and system components.

Occupational Health and Safety Act—Regulations for Construction Projects
(Each province has their own version, so you will have to acquire the one for your province.)
Covers minimum standards for safety on construction sites.

Notes